T0349159

PORTRAIT
OF AN OYSTER

ANDREAS AMMER

Translated by RENÉE VON PASCHEN

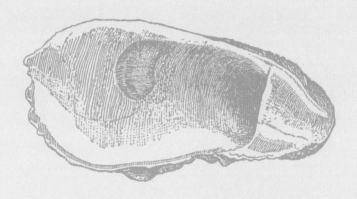

PORTRAIT

— *of an* —

OYSTER

A Natural History of
an Epicurean Delight

GREYSTONE BOOKS
Vancouver/Berkeley/London

Greystone Books Ltd.
greystonebooks.com

Cataloguing data available from Library and Archives Canada
ISBN 978-1-77840-127-5 (cloth)
ISBN 978-1-77840-128-2 (epub)

Editing for English edition by James Penco
Proofreading by Alison Strobel
Indexing by Stephen Ullstrom
Scientific review by Dr. Matthew W. Gray
Cover design by Javana Boothe
Cover illustration by George Shaw, *The Denticulated Oyster*
Text design by Fiona Siu

Printed and bound in China on FSC® certified paper at Shenzhen Reliance
Printing. The FSC® label means that materials used for the product have
been responsibly sourced.

Greystone Books thanks the Canada Council for the Arts, the British Columbia
Arts Council, the Province of British Columbia through the Book Publishing Tax
Credit, and the Government of Canada for supporting our publishing activities.

Canadä

Greystone Books gratefully acknowledges the xʷməθkʷəy̓əm (Musqueam),
Skwx̱wú7mesh (Squamish), and səlilwətaɬ (Tsleil-Waututh) peoples on
whose land our Vancouver head office is located.

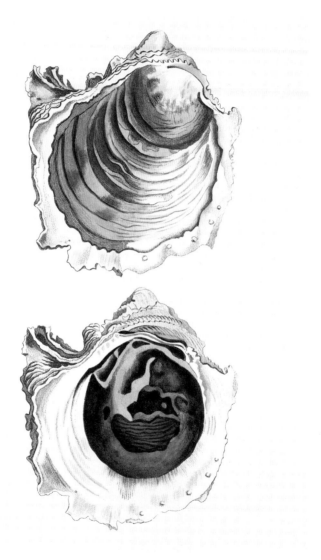

CONTENTS

PORTRAITS

THE PHILOSOPHER'S MOLLUSKS

> "And many internal things in man are like the oyster—
> repulsive, slippery, and hard to grasp."
> FRIEDRICH NIETZSCHE, *Thus Spake Zarathustra*

THERE IS A PAINTING by Édouard Manet called *Beggar With Oysters (Philosopher)*. A beggar is portrayed in it, looking at us sadly, as though he were about to say something but prefers to withhold it instead. He realizes that we might not listen to his words of wisdom on account of his shabby appearance. He has an azure shawl wrapped over his ragged, taupe-colored attire. A street scene, perhaps—however, we can only presume. The painting has neither background, nor setting, nor does it give any indication of the milieu. Thus, nothing can detract from the only prop in the picture, which does provide a clue: at the lower right-hand edge of the painting, before the beggar-philosopher's feet, there are half a dozen oysters—bedded on a small pile of

A coincidental meeting between two forms of life, of which only one will survive. Édouard Manet, Beggar With Oysters (Philosopher), circa 1865.

straw. Their shells reflect the color of the man's rags. The majority of the motionless shellfish are closed. Yet one is open and so bright that its flesh outshines the primarily darker shades of the painting. That oyster is also—as to be expected—taciturn.

The picture could be interpreted with a saying like "tough on the outside, but soft on the inside," a warning not to judge a book by its cover—much like the rough-looking, unassuming outer appearance of a being that may encase something more cherished. However, in the middle of the nineteenth century, when Manet was completing his picture of the oyster-philosopher, the oyster was not necessarily a sign of luxury and decadence. Instead, it was sold in the street to the masses as a cheap form of nourishment, which humanity greedily consumed almost to the point of extinction. Akin to Hamlet and the skull, *Beggar With Oysters* reflects a pearl of wisdom about the state of the world and humankind, found while contemplating an oyster on the half shell.

> "First I ordered a dozen oysters."
> **WALTER BENJAMIN**, "Hashish in Marseilles"

The task of writing a "portrait" of the oyster involves a problem of almost quantum-mechanical dimensions, as it is practically impossible to observe a living, uninjured oyster outside of its shell or in its natural habitat. The first encounter between the creature and a human being typically takes place at a seafood market. According to the law, at least in Germany where I live, each wooden crate of oysters must have a packaging label affixed to it declaring: "Contains Mollusks. These Shellfish Must Be Sold Live." Living within are silent, immobile creatures, until their incredibly hard shells are forcibly shucked for consumption, revealing a soft, now-unprotected creature. No one who has ever eaten an oyster has experienced this magical moment without a certain sense of awe.

Before me is a plate of crushed ice with a dozen oysters on the half shell. After they have been shucked, the same is true as before: the oyster should still be alive. However, it is a creature awaiting its death after its powerful muscle has been shorn off. Proof of this can be seen when the creature before me contracts upon touch, or if I squeeze a few drops of lemon juice onto it— the only sign of life I can see. Yet we probably should not construe this reaction as friendly greeting.

I love oysters. Shucking them with an effort; slurping the cool, slippery form within. Holding them on my tongue for a moment before chewing a few times; finally experiencing the unique, clear taste of the sea. They are comparable to nothing else in the world. The enjoyment is always too brief.

Oysters are a poetic dish, yet simultaneously an existential one. Consuming an oyster involves an encounter of two living creatures—although it is neither a visual nor philosophical experience for the oyster, since it lacks eyes and is not capable of thought as we know it. The upper, so-called right-hand side of the oyster shell can only be removed with brute strength using a shucking knife and acquired technique, along with a certain amount of finesse. With the help of the law of the lever, the two halves of the shell are forced apart at a certain point that is only learned by practice. The muscle, which the little creature employs with all its might to resist its demise, is simply cut apart. Without the requisite tool and know-how, most people would not be able to solve the puzzle of shucking the oyster. In achieving the feat, we reduce ourselves to the level of a creature for a moment: neither an egg, nor a rare steak, nor a salad so innately remind us of the archaic laws of nature, which state that human beings, with flesh-tearing incisors, are also predators that have to kill their prey before consuming it—armed with a knife, if need be.

The wound inflicted on the muscle by the shucking knife is the sole culinary technique required, as a rule. Preparing an oyster, apart from the complicated art of *haute cuisine,* only entails opening its shell. It is eaten as is and while alive. To complement the sensory aftertaste in the mouth, a chilled drink is recommended. I take a sip.

The second oyster lies glistening and moist in my hand. *Videri est morire* is the maxim in the case of the oyster: "To be seen is to die." Only in its final moment of life does the oyster no longer present its shell or outer enclosure to me, once its interior has become visible as a living creature. Naked, immobile, and silent, it just lies there. Does it know any fear at that moment, or does it experience a moment of epiphany, when it encounters me, another form of life, a more intelligent form? It is the first time an oyster finds itself unprotected in the world, without its shell. A moment later, I swallow it.

I pick up the next one. When the living creatures of human being and oyster encounter each other, the latter is incapable of moving. It cannot run away. It is a mollusk, blind and purportedly also mute, one of the earliest forms of life. It is faced with human beings, used to considering themselves a higher form of life. An oyster farmer, one of my contemporaries, did everything he could to give his oysters the best possible living conditions so they would grow and thrive. He worked long and hard for the moment during which I experience a few seconds of culinary delight. To be or not to be? Pleasure or regret? Swallow or chew? Although the oyster remains silent, once we begin contemplating the humble form of life in our hand, we are thinking primarily of ourselves. Tales of oysters are tales of people. Whosoever contemplates an oyster, contemplates themselves—until they swallow it.

Those few seconds, during which the oyster lies in my hand doomed between life and death, constitutes a heavenly moment. I have a premonition of the ocean, from which we all evolved, which soon evokes a pleasant, salty, primeval sensation. The moment between its existence as a mere creature and as the peak of culinary delight is a great pinnacle of human experience. In this moment, I am part predator; part ancient god; dependent on nutrition, with a taste for pleasure, yet capable of thought, empathy, and a guilty conscience.

Although the slippery shellfish in front of me is a creature, it would not be easy to find in a zoo. The life of an oyster, although it can be grown in aquaculture, is hidden from humankind, just like that of a pig on an industrial-scale farm. However, in contrast to all the other creatures slaughtered for us, we are personally responsible for the death of the oyster. No one else takes the fatal bite for us.

Six empty shells of consumed creatures are lying upside down on the crushed ice. Half a dozen are still waiting, their soft, beige, and silvery shimmering bodies on the half shell before me. I might use my fork to detach their bodies from the shell—if the professional oyster shucker has not already done so—then, I'll squeeze a little lemon juice on their dark outer extremities, looking forward to the moment when they contract. The contraction is considered a sure sign that the creature in hand is still alive. Sensitive people, who prefer to have their food slaughtered by someone else, are turned off by this. Woody Allen says in his comedy *Don't Drink the Water*, "Oysters are alive. I don't eat live food. I want my food dead—not sick, not wounded—dead." From a rational perspective, it's not quite clear what this preference for dead creatures is based upon. After all, humans are omnivores—not vultures.

I'm sure that if the oyster had an eye upon me before I bit into it, I wouldn't enjoy eating it. The moment before I open my mouth to swallow the live oyster, I cannot feel it observing me. I cannot imagine the oyster "knows" me. At that moment, a person may feel like a primeval god, or at least like something unfathomable to the oyster. The creature lacks the categories to perceive us—as Immanuel Kant, the oyster lover, would have said. Just as an oyster can't read Kant, likewise it can't sense fear. And I gulp down the live oyster in a single bite.

The relationship between humans and oysters is unique, as most people's only interaction with the creature takes places in the mouth via our sense of taste. It can be considered almost an erotic act, as well. Our tongue plays around with the oyster for longer than we take to look upon it. Most of the zoological and artistic portrayals of oysters are depicted only after their muscle has been severed by force, and the upper half of the shell has been removed. Strictly speaking, we do not know what an unopened oyster looks like. If oysters had red blood, no one would have considered eating them raw. Do oysters actually have any blood? Zoologists have at least discovered a tiny heart. The oyster's heart, sometimes visible beating next to its adductor muscle, consists of three small chambers. However, it only pumps a milky substance—with oxygen absorbed from the seawater—through the creature's open circulatory system. This system is filled entirely with salt water, as the oyster filters pure seawater through its body. That is how the so-called oyster liquor—considered an actual delicacy, the creature's "nectar" with a particularly refined taste—is immediately formed, when the water in the shell is poured away after being shucked. Human beings might not only be predators, but even blood-suckers on occasion.

I help myself to another oyster. It would be an exaggeration to say that an oyster has any kind of beauty—on the contrary, anyone who has ever seen an oyster before it's been industrially cleansed inside and out needs a lot of know-how to be able to consume it, as it is probably encased in a slimy layer of decaying algae and mud. This unappetizing, moldering, irregularly shaped clump has little in common with an internationally marketed delicacy. The origin of the oyster in the slimy mud of estuaries and seacoasts is comparable to an abattoir—a taboo zone that is normally excluded from the human experience of

All that remains after the meal: the hard shell of an oyster with its gleaming inside.

gathering food. The savvy connoisseur, dining at a table with a lily-white tablecloth, does not want to know anything about it.

Presumably, the oyster dies a sudden death at the exact moment that I bite into it—an action that not everyone does, although it heightens the culinary experience. It is an archaic act that takes place in a fancy restaurant, the act that life has been based upon for millions of years: the moment when one animal bites another in the neck for the purpose of killing and devouring it. As a human, my only experience of such a moment is when I eat an oyster. My bite turns me into the lion that kills a gazelle after a wild hunt, securing my continued survival!

Despite all my research, I do not know exactly when the oyster becomes clinically dead. Sometime after I sever its powerful muscle? Perhaps when I injure its heart, next to the muscle, while shucking it? Then the creature would already be dead, and the final contraction would only be a chemical reaction rather than its final sign of life. It could even be a postmortal electrical reaction, such as that of frogs or fishes, whose cadavers quiver after they have expired. Or is it a desperate call for help? Instead of using a knife, I could use the ancient technique of fire, as did Stone Age people. If the oyster is heated up, its point of death can be precisely determined. It suddenly gives up its life force. Its powerful muscle weakens, and the oyster opens mechanically all by itself.

However, if I open the side of the oyster using a shucking knife, the creature should still be alive. It does not have any premonition of my dark maw that will devour it; or my tongue and esophagus, which it will slide down; or my stomach, where it may land in a pool of champagne. Might it survive a moment longer if I don't bite into it?

I slurp the last oyster from its shell with gusto. I have not chosen particularly large oysters, because the smaller ones often taste better and do not fill the mouth as much. Disregarding the advice of connoisseurs, I do not chew it this time. I slurp its so-called oyster liquor, and I am arbitrarily reminded of the time when all creatures could taste the fresh, salty sea, which was the primeval origin of humankind and all other life on Earth. The creature survived that long journey to me in this tiny pool of water. Did it enjoy the moment spent in my mouth, touched by another living creature, gently and warmly caressed by my tongue? Or was the tiny creature in my mouth afraid of its demise? Do oysters have feelings they cannot show? Does each culinary delight also mean destruction? Or does a person simply give it too much thought when writing a book about oysters?

Irregularly formed creatures for perfect moments.

THE PERFECT
MOMENT

"He was a bold man, that first [ate] an oyster."
JONATHAN SWIFT, *A Complete Collection of Genteel and Ingenious Conversation*

HARDLY ANYONE EATS OYSTERS nowadays for the sole purpose of sating their hunger. They only serve the purpose of nutrition to a limited degree. Consuming them constitutes a magical moment rather than creaturely necessity, a sensual experience rather than fulfillment of a basic need. In that sense, oysters now have more in common with art than with food. They represent enjoyment, or the premonition of it. "It's a one-on-one relationship," according to the greatest living expert on oysters, the American Rowan Jacobsen, who has tasted and described each of the over three hundred varieties found in North America. The oyster expert has pronounced dining on a dozen oysters to be one of the last remaining cultural rites of the animal sacrifice, a "Bronze Age atavism" still practiced today at the marble altars of the oyster "temples" in New Orleans, on the coasts of France, or in our own dining rooms, whenever an oyster is shucked:

> The high priest greets you; ritual conversation ensues. Then he raises his knife and cuts the muscles of a dozen oysters as you follow his clean, rehearsed motions with your eyes: *Hoc est enim Corpus meum, quod pro vobis tradetur.* He sets the offering before you, you anoint it, and the deed is done. Wine is splashed, a little tithing for the priest, the gods are pleased, and the universe has been renewed for another day.

As he says, we know two kinds of sensual enjoyment. On the one hand, immediate fulfillment, such as ice cream, sex, and drugs. However, there are others that require due time and energy to fully enjoy them, including poetry, cooking, and physical exercise. Once the effort is over, we have a sense of satisfaction and feeling alive. Eating oysters belongs to the second category; it hones the senses and may even become a life-changing experience. Generally, everyone recalls eating their first oyster, just like they remember their first kiss—yet not their first carrot.

It overcomes you unexpectedly like a brain wave. A short and sweet moment that anyone might experience: the moment that can suddenly and unexpectedly change everything in life. The magical moment after which nothing will be the same.

"Everything was different now." Those four words summarize the reaction of one of the most famous gourmets on this planet after his first encounter with a raw oyster: Anthony Bourdain, who first became a famous cook, and later an even more famous television star. As a great gourmet, he traveled the globe making television presentations about the dishes served to him on his trips and the beauty of the world. His television shows *A Cook's Tour* and *No Reservations* constituted the invention of some of the first successful gourmet travel shows. In his autobiography *Kitchen Confidential*, he writes that he spent the summer holidays with his parents in France, where he heard "A Whiter Shade of Pale" on the jukeboxes and was invited by his aunt's neighbors to an excursion on an oyster barge. After he had eaten a baguette with cheese, young Anthony was still hungry, so the oyster fisher Monsieur Saint-Jour offered him an oyster. Anthony's parents hesitated, and his brother politely refused. "But I, in the proudest moment of my young life, stood up smartly, grinning with defiance, and volunteered to be the first." And then it happened:

> And in that unforgettably sweet moment in my personal history, that one moment still more alive for me than so many of the other "firsts" which followed—first pussy, first joint, first day in high school, first published book, or any other thing—I attained glory.

The oyster fisher shucked a "huge and irregularly shaped" oyster. The younger brother was aghast at the slippery, ostensibly sexual-looking object that was alive and dripping. Little Anthony, however, "took it in [his] hand, tilted the shell back into [his] mouth as instructed by the by now beaming Monsieur Saint-Jour, and with one bite and a slurp, wolfed it down. It tasted of seawater . . . of brine and flesh . . . and somehow . . . of the future." And then it happened. Bourdain, the future world star, immediately knew that his life had changed. "Everything was different now. Everything." And when he wrote his memoirs, he admitted:

> I had, somehow, become a man. I had had an *adventure*, tasted forbidden fruit and everything that followed in my life—the food, the long and often stupid and self-destructive chase for *the next thing*, whether it was drugs or sex or some other new sensation—would all stem from this moment.

Bourdain is certainly not the only one to have had such an experience. At the seashore, I recall my father showing me how to take a raw shellfish from the stand of the fishmonger, who immediately shucked it with a knowing smile. Next, they were both able to observe that familiar play of emotions—fear and courage, triumph and survival—in my expression. I am still convinced I can recall that the fisherman's wooden cart was blue.

The other "first oyster" I ate was at the beginning of my adulthood, during my first trip with the love of my life on the coast of the Mont Saint-Michel Bay in Cancale. After her attempted protest, I bought us a dozen oysters and took them up to our tiny hotel room. As a young, unmarried couple, we hadn't been given a room together in Brussels, but here the hotel manager had lent us a shucking knife. The love of my life had never eaten an oyster. "Just shared one with my mother," she claimed, not realizing that that was a next-to-impossible feat. And I can swear that in one of those first twelve oysters we

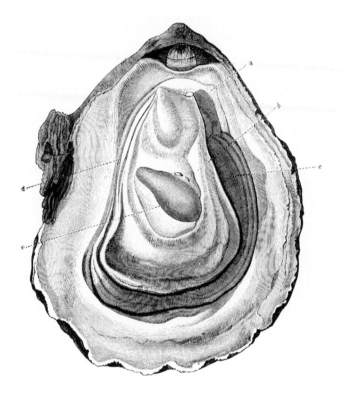

An erotic image from Brehm's Life of Animals? *The soft body of the oyster on the half shell is an animalistic depiction of the death of childhood.*

ate together, we found a flat, shiny object that we decided to call a pearl, although we knew that edible oysters do not normally produce any pearls. Nevertheless, we both saw it. However, we cannot prove it anymore, since the pearl was lost long ago—yet not the memory of that evening.

Rowan Jacobsen, in his writing on the oyster, has coined an expression for the phenomenon of becoming an adult while eating oysters. He calls it the "Oyster Conversion Experience."

> The Oyster Conversion Experience is remarkably consistent among individuals, genders, and generations. You are an adolescent. You are in the company of adults, among whom you desperately want to be accepted. You are presented with an oyster, you overcome your initial fear or revulsion, take the plunge, and afterward feel brave and proud and relieved. You want to do it again.

It is true, however, that not all people are able to envision their future, the love of their life, or their adulthood the first time they consume an oyster. As a source of nourishment, the oyster divides humankind. There is not really a gray zone between oyster lovers and oyster haters. Either people enjoy the shellfish at any opportunity that presents itself, or they find them disgusting. Oyster haters often claim they are tasteless or slippery. They may well be slippery, but the same can be said of avocados and egg yolks. Overcoming the sensation of slipperiness is part of the enjoyment. And of course, the sensation of enjoyment has an erotic component. That is likely why the scene described by Anthony Bourdain takes place on the threshold to adulthood. Oysters are meant for adults.

"An oyster," writes Rowan Jacobsen, "like a lover, first captures you by bewitching your mind." An oyster first entices your fantasies, like any true love. Their wet, slippery consistency is reminiscent of sex organs or—although this is not always perceived with pleasure—of penetrating a foreign body. The biographer Andrew Wilson, in his book *Beautiful Shadow,* recounts the story of sixteen year-old Patricia Highsmith, who was completely turned off by the farewell kiss her male companion gave her one evening. The future crime writer found the experience of kissing a man, to whom she was not particularly attracted, to be like "falling into a bucket of oysters."

No other food has been attributed with such an erotic lure. "Think of oysters, think of sex," writes Rebecca Stott in her book *Oyster,* one of the most interesting books about the animal. That's not only connected to the supposed similarity of the wet marine creature to a vulva. Everything about an oyster can potentially be erotic.

It is not particularly important whether oysters fire the erotic drive physically or chemically. Their cultural attributions of strength, potency, and delicacy have given rise to the strangest manifestations. Even the word "aphrodisiac" is connected to this marine creature, since Aphrodite, the goddess of love, washed up on shore in a seashell in Greek mythology.

With little credibility, the high zinc content found in oysters is often seen as responsible for the stimulating effect of oysters. Zinc does play a limited role in the production of the male sex hormone testosterone; however, an ejaculation of semen only contains a couple of milligrams of zinc. A single oyster already contains far more of this metal than required for such a minute volume. For years, the world was waiting for scientific proof as to whether the myth had a grain of truth. In 2005, research was finally presented by Italian and American

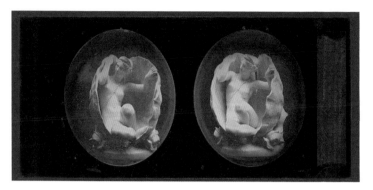

An immersive work of art: daguerreotype of the sculpture Venus With a Shell by James Pradier.

scientists at an American Chemical Society meeting in San Diego. Raul Mirza and his colleagues claimed that the amino acids known as D-aspartic acid and N-methyl-D-aspartate, found in oysters, mussels, and clams, enhance the production of testosterone, as well as the female sex hormone progesterone. The scientists claimed, "Yes, we do think these mollusks are aphrodisiacs." They were sure to make headlines, yet they did not garner applause from their colleagues in science. Nothing has been proven to this day.

Nevertheless, the culinary enjoyment is very real. We relish eating these shellfish, as long as we are not disgusted by them as by an unwelcome display of sex organs. Oysters are not eaten like chips or peanuts; rather, they are eaten with care, more or less tenderly. Although oysters cannot jump out of their shells, we approach them like a hunter or a lover, with due care and observation, watching for each quiver and sign of life. We salve them with a few drops of lemon juice. We slurp them directly from the shell, gently chewing the cold mass, before putting the remaining shell back onto the plate, with a final glance at the mother-of-pearl. The half shell is turned upside down, like closing the eyes of the deceased.

The lesson of the meal: enjoy the moment—*carpe diem*—although it may be as fleeting as the bite of an oyster, which has done nothing much in life besides gorging and multiplying itself.

TWO GENDERS?

"I had rather be an oyster than a man, the most
stupid and senseless of animals than a reasonable
mind tortured with an extreme innate desire of
that perfection which it despairs to obtain."

GEORGE BERKELEY

CAN OYSTERS BE HAPPY? Do they dream of swaying in the gentle waves of the sea? At least they are easy to please. They do not pine for distant shores; instead, they live contentedly in the very same place they chose as home in their early youth. Their lives are nevertheless unfettered, as they consist of truly little apart from feeding themselves and reproducing. Both are solitary activities for them, including sex. An oyster does not fall in love. Its propagation almost seems like euphoric wastefulness, an unoriented flow of life in the watery universe of the primeval ocean. As such, the oyster can be perceived as a creature that is content to remain single for a lifetime. It cannot exist outside of its hard shell.

In fact, it is extremely unlikely that an oyster sperm would just happen to find an oyster egg as it swam in the almost infinite vastness of the ocean. As M.F.K. Fisher explains in her epochal book *Consider the Oyster*, each new oyster has been propagated in nature thanks to a "potent and unknown sire" and a most patient mother. A young oyster will never meet their male propagator—and rarely their female counterpart. The parents have no interest whatsoever in their children. The father oyster spawns sperm into the sea as a lonely bachelor. With a stroke of luck, the sperm might find its way to a female oyster or one of her eggs—depending on the species, each female can spawn around a hundred million eggs (but who has counted them all?)—which are released by the mother oyster shortly after they are fertilized. The immense ocean is the mollusk's nursery.

There is a certain chance that the oyster father is younger and less experienced than the oyster mother in this disinterested act of reproduction. The reason being that oysters undergo gender transformation several times during the course of their lives. In the case of the Pacific oyster, which is the most widely farmed globally, the spat first assume the male gender. The following year, these creatures undergo transformation into females, and then back to males again—according to whim, it might seem to human beings. They might be called "gender-fluid" in the terminology of the contemporary LGBTQ+ community. The scientific term is "sequential hermaphroditism." The gender transformation itself takes around two weeks. The gonad, the sex organ located next to the oyster's heart, can produce either sperm or eggs, one after the other, a feat which is difficult for us to comprehend. For the oyster, the transformation from female to male is faster and less complicated

S.T.A.I., La Perle, anonymous postcard, circa 1930.

than from male to female. Why and how this takes place is not entirely clear. In colder climates and waters, such as along the coastline of Great Britain, gender transformation only takes place once a year. However, it may take place more frequently in the Mediterranean or the Bay of Biscay. Some researchers hypothesize that the female is so depleted after having produced millions of eggs that she transforms herself into a male to regenerate her strength. The more abundant the source of food, the more likely there will be a large proportion of females in a population. Yet if conditions deteriorate, then the less-demanding male gender will become predominant to guarantee survival. On the whole, it would seem more attractive to live the life of a female oyster. The longer an oyster lives, the more likely it is to take on the female gender again.

The oyster's love life has not yet been entirely clarified, so we do not exactly know how oysters attract one another, how they communicate while living apart from each other in their separate shells, or how they propagate themselves—simultaneously spawning millions of eggs or sperm into the sea. It appears, however, that although they may live far apart from each other on cliffs and stones, or attached to other shellfish, they are able to stimulate each other hormonally, despite the distance. As soon as the water temperatures surpass a stable level of 64 degrees Fahrenheit, these creatures begin to feel good. Soon the first male oysters begin spawning on their own, which might seem like a homoerotic act. They say that male oysters that have come into contact with the sperm of their contemporaries are more likely to demonstrate increased reproductive activity. On such days, the seawater can even turn milky with oyster sperm, in areas where oysters are spawning. The free-floating sperm somehow manages, although statistically improbable, to reach the output of the female oysters in sufficient volumes, fertilizing their millions of ripe eggs.

In the seas of our day and age, this unique method of fertilization rarely takes place any longer. Wild oysters have almost become a quirk of nature. The warm tidal flats of Arcachon, west of Bordeaux, are one region where oysters can multiply— closely cared for by oyster farmers—outside of huge glass tanks. Instead, oysters are bred almost exclusively in gigantic aquariums. Biologists observe when the male oysters begin spawning. The water from their tank is then mixed with the water of the female oyster tank to achieve an optimal ratio between oyster sperm and eggs. It does not require a great deal of masculine prowess, as half a teaspoon of male oyster spawn can fertilize several million eggs. A more brutal method involves the removal of the male sex organs from the oyster with tweezers. They are then put in a mixer and blended with an exact ratio of oyster eggs to produce the larvae. In North America, and even in the great oyster farming regions in France, the oyster farmers are already almost entirely dependent on oyster farms. Hatcheries deliver larvae, which are deposited in tanks, which can then be planted in aquaculture or fisheries a week later, in a process known as remote setting. When the seed are introduced to the sea, they are around two to three millimeters in size.

When oysters do multiply at sea, their larvae are able to float by themselves for approximately two weeks after fertilization, when they are known as veliger larvae. By then, they are actually still at the stage of an egg, in biological terms. M.F.K. Fisher writes in *Consider the Oyster* that this is the only period in the life of an oyster when it may enjoy itself. "It is to be hoped, sentimentally at least, that the spat—our spat—enjoys himself. Those two weeks are his one taste of vagabondage, of devil-may-care free roaming."

That is the end of the oyster's free-floating life, until its probable death upon consumption by a human being or another predator from the animal world. The planktonic stage of the oyster's life, in its first few hours before forming a shell, is its most vulnerable stage, comparable to that of puberty in humankind. Oyster larvae constitute tasty food for most other marine creatures, from whales to coral—as well as their own parents.

Let us assume that one of the million oyster eggs survives its early youth, despite all odds. Before taking up permanent residence somewhere for the rest of its life, the tiny creature undergoes a short stage of having a limited power of vision via a primitive organ, enabling it only to differentiate between light and dark, which disappears again during adolescence. With the help of a foot, which the oyster also has during this early stage and then loses soon after, the tiny creature can move around a little to find a better place to live. Yet its promenades will not last long.

After a few weeks of swimming around in the sea, the creature, measuring around one-third of a millimeter at this point, will attach its "left foot"—the thicker bottom half of the oyster's asymmetrical shell—with the help of a cement-like glue to the

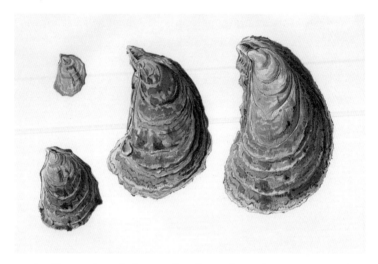

Oysters through the ages (smallest to largest): one, two, six, and eight years old. Sherman Foote Denton, Oysters, Natural Size, 1902.

substrate of its choice, known as a cultch. Biologists have calculated that only around 250 out of 1 million eggs reach the adolescent stage of a spat. And only a dozen of these survives their first winter. That means the twelve oysters on my plate are the only survivors out of several million siblings that were conceived.

From the moment it becomes a permanent resident, the tiny oyster immediately "devotes himself to drinking," according to M.F.K. Fisher. The oyster digests the contents of its watery environment in the literal sense. The underwater world constitutes a real smorgasbord. Its meals are readily available. The oyster does not need to move, as the nutrition floats right through it.

Most oysters in natural conditions filter about half a gallon of water per hour, or about ten to twenty gallons of water each day. It flushes the seawater through its body and filters many things out of the water, from plankton to other oysters' eggs, as well as human excrement. After one hundred days, the creature, or its shell, may have already grown to a size of 1.5 to 2 inches. During its permanent feeding frenzy, the oyster can undergo gender transformation according to whim—and if an oyster fisher or farmer does not collect it—then it may grow to an age of ten to twenty years and a weight of over four pounds.

Those oysters served to us in restaurants were not free to choose their place of residence in the sea on their own. If they were to continue their lives in one of the oyster farmer's plastic bags, you might say they already belonged to the jet set of their species—the most immobile of all creatures is carried around by oyster farmers according to need. No matter whether an oyster was conceived at sea by its parents or in a plastic tank by biologists, it will spend the rest of its life together with its contemporaries in the wildly romantic straits along the coast, with the tides constantly washing across it, cleansed and turned over during ebb tide by the hardworking oyster farmers. They usually pack the mollusks in sacks placed in the water on tables, so they will not be covered in mud or devoured by predators. The sacks are regularly turned over at ebb tide, ensuring that the shellfish won't stick together, instead taking on the individual drop-formed shape valued in restaurants around the world. At the end of its life, after three years on an oyster farm, a European mollusk may be able to spend its final weeks in a comfortably

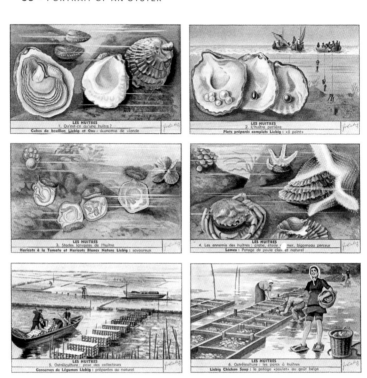

French collectible cards with a philosophical riddle:
"What is an oyster?"

warm pond of tasty brackish water in the French region of
Marennes-Oléron. Finally, it is cleansed and undergoes depuration, refining its taste, before it is bedded in algae and packed in
wooden boxes in refrigerated trucks or airplanes to be shipped
around the world. No matter their destination, oysters are sure
to be presented in high-end shops or upscale gourmet restaurants where people may not always find it easy to get a table.

Oysters have a heart that pumps, but not a brain that thinks. They have a mouth, a relatively complicated stomach, intestines, and an anus, as well as gills. Tiny cilia are found on the gills, which create a kind of suction, making the surrounding water flow through the gills. These tiny cilia react to anything acidic with a spasm. But what does an oyster eat? Scientists have thoroughly investigated the complex form of nutrition of the oyster. The description of this process by Berlin zoologist Rudolf Kilias is not necessarily appetizing:

> The stream of water passing through the mollusk first enters the ventral (convex) side of the mantle, passing from there to the gills on the dorsal (back) side, and is then expelled from the mantle near the anus through the so-called excurrent siphon. The particles remaining caught in the mucus of the gills are drawn along the cilia towards the mouth by the labial palps on both sides of it, which regularly transport the nutrients into its mouth.

However, an oyster does not consider everything appealing. It carefully sorts what it finds in the seawater:

> The nondigestible substances are then transported via other cilia to the vicinity of the excurrent siphon, where they are stored at first. As the shellfish regularly opens and closes its shell, it expels these unsuitable particles (pseudofeces) along with the water and its feces.

This complicated digestive system is not always in use. After a severe winter, the intestine is usually empty; however, it will be well filled during the spawning period beginning in July. The oyster's metabolism is so active during the summer that oyster beds need to be continually cleansed of oyster droppings, like horse stalls.

The oyster's most powerful organ is its adductor muscle—a body part that would enable a four-inch-large shellfish to easily lift a weight of twenty pounds, if it weren't using all its force to clamp shut the asymmetrical halves of its shell whenever needed. The oyster only has that one single muscle. The only binary digital motion this mono-muscular creature is capable of is opening and closing its shell—whereby opening it is involuntary. Open or shut. Evolution figured it out in the case of the oyster; it doesn't need a brain to control its muscle. A handful of chemical processes and a homemade protective shield are completely sufficient for leading a fulfilled life.

Half of the oyster's muscle strands are geared toward speed, ensuring the shell can close quickly in case of danger. These do not have much tenacity and cannot keep the shell closed for long. This important function, enabling the oyster to survive being stranded by low tide, is performed by the darker part of the muscle. Without severing this muscle, it is impossible for a person to shuck the delicacy. It is the lasting force of this muscle that ensures the shellfish will be alive after their days-long transportation. The muscle keeps the oyster closed—and it is actually trained by the oyster farmers. The marine creatures are repeatedly lowered into the water of the depuration tanks, where they open, and then left stranded to dry, where they close tightly with a sufficient supply of salt water. Before they are

The mollusk ... not a lot more than its own lifeless shell, which will long outlive the creature that created it.

shipped, the time they spend high and dry is extended. In the end, an oyster can survive for up to two weeks out of the water, as it can emit water itself; this is also the maximum period that it can be shipped for delivery. An open oyster without any water is a dead oyster. And a dead oyster is not a good oyster.

Adult oysters have multifunctional sexual organs—gonads—however they lack any sensory organs. Nevertheless, they can perceive their environment. If there is a shortage of water, no matter whether the tide has ebbed or the oyster is being shipped, it closes its shell. It will react to light and shade, as well as the salt content of the water—or if you knock on its shell. It is said that the oyster has two senses, in contrast to humankind's five senses. To be more concise, it only has a few simple pressure sensors. However, the creature's limited sense of feeling and its nonexistent intelligence have enabled the species to survive for approximately 250 million years. On the other hand, even if one muscle and a few pressure sensors are sufficient for survival, they do not enable the creature to recognize human beings as deadly foes (or as oyster breeders).

No matter how long an oyster lives, it actually leads a heavenly life. Contented like Buddha and literally resting at ease, it drinks as much as possible, fanning nutrition through its body, and simply shuts its powerful shell should anything bother it. Its natural enemies are limited. In order to threaten an oyster, a creature needs to be specialized in terms of evolutionary biology. Several kinds of starfish can wrap their arms around an oyster, opening its shells a crack, and force their own stomach into its shell. The starfish then digests the contents of the oyster

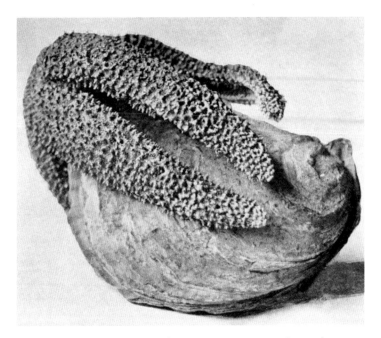

The greatest competitor for oysters: in contrast to human beings, starfish have found a means to open oysters without using tools.

shell from outside. Naturally, any creature would be defenseless against a treacherous attack like that, which could have come from one of the *Alien* movies. Another of the oyster's natural enemies—the oyster drill—is so highly specialized in feeding on oysters that it was even named after its prey. With the help of an acid, this sea snail drills holes in the oyster's shell and sucks it out through the hole. Other common predators of oysters include crabs and seabirds—most obviously, the oystercatcher—as well as sea otters, which hit pairs of oyster shells together or against rocks to crack them open. Human beings, who began eating oysters after the earliest stages of our evolution, do not possess such a fine-honed technique. In the fair contest of creature against creature, humankind often loses. Without a handy tool, a person cannot shuck an oyster. Nevertheless, in a show of greed, human beings have almost managed to drive oysters into extinction twice before discovering a more productive role as steward, farmer, merchant, and conservationist.

*W.G. Mason, The Oyster Stall, in Henry Mayhew's
London Labour and the London Poor, 1851.*

BREEDING AND CLASSIFICATION

> "The first person who formed artificial oyster-beds
> was Sergius Orata, who established them at Baiae,
> in the time of L. Crassus, the orator, just before the
> Marsic War. This was done by him, not for the gratification
> of gluttony, but of avarice, as he contrived to make a
> large income by the exercise of his ingenuity."
> **PLINY THE ELDER**, *Natural History*

AS THE PLENITUDE OF OYSTERS began diminishing in the ancient world, the Greeks began distributing terra-cotta shards along the coastline of the sea, as they knew that oysters preferably spent their lives attached to terra-cotta. A little later in history, when the Roman nobility had consumed practically all the oysters in the Mediterranean, a certain Gaius Sergius Orata managed to cultivate an especially beautiful characteristic known as "calliblephara" (meaning "with beautiful eyebrows") because of the thin purple thread running along the inside of the oyster. He is mentioned in this book not because he is also said to have invented a form of floor heating, but because he devised a way to easily transplant oysters by means of twigs distributed in the vicinity of adult oysters, which then captured their spat. He then farmed the oysters in Lucrine Lake, which he cleverly supplied with seawater via a system of canals from the Gulf of Naples. His customers were the richest Romans; Orata soon became one of them.

Curiously, his technique of oyster cultivation was then forgotten for almost two thousand years. It was not until the nineteenth century, when wild oysters had almost become extinct once again—although they had been considered plentiful—that the technique of oyster cultivation was finally reinvented. In France, the oyster-loving Emperor Napoleon III personally commanded that methods be sought for cultivating oysters, which could until then only be harvested in the wild—ships had been dredging oysters by the millions from the diversely inhabited seabed.

Once all the oysters had been dredged, the biggest problem in oyster farming was how to gather the plankton-sized oyster larvae. In the middle of the nineteenth century, the inspector general of maritime fishing, Jean Victor Coste, who had been appointed by the emperor, began using stacked terra-cotta roof tiles to lure the larvae to attach themselves, a system that had already been employed in Imperial Rome. In 1865, a mason by the name of Jean Michelet suggested that the state oyster farming guild could apply a layer of lime to the terra-cotta tiles, making it easier to detach the spat from the tiles for transportation in bags. If you visit the French oyster farmers at Arcachon

Oyster dredging on the French coast in times past. After a few years, the entire population of oysters had almost become extinct, although they had been considered infinitely plentiful. That marked the onset of oyster farming, which had not been needed in the past.

today, they will show you the white-painted clay tiles known as *collecteurs* to demonstrate how the spat are traditionally gathered. Experience has shown that several plastic boxes stacked on top of each other can also serve the same purpose. The spat, which measure several millimeters in size, are brushed off after a few months, washed, and shipped around the world. Nowadays, hardly any oyster farms still maintain their own nursery.

Many oysters farmed in France spend their juvenile years in Arcachon, their adult lives in Brittany, and their old age in Marennes-Oléron, where they are cleansed and undergo depuration. In England, the final process of depuration is left out.

There, oysters only require cleansing for twenty-four hours in water sterilized by UV rays.

In many gourmet shops, farmed oysters from France are sold as *fines de claire.* In this context, *claire* refers to a shallow salt pond where the oysters reside for a few weeks for depuration. These ponds are located primarily in the poorly accessible tidal flats of the Seudre River, which is flooded daily by saltwater tides and appears strangely deserted during the daytime. Those preferring more highly refined oysters should buy the more expensive *spéciales de claire.* These shellfish undergo a depuration process over two months, which means they are more highly "purified"; however, they are also somewhat blander.

Up to this point, we have been describing oysters that can be compared to sparkling wine, according to the French designations. If a person prefers champagne, they should obtain oysters from Marennes-Oléron, the only oyster farming region in the world with a "protected geographical indication" (PGI). Almost one-quarter of the French production, comprising around thirty thousand tons of oysters, are produced there annually. There are few landscapes that look *less* impressive than this incredible gourmet region. In this region in the west of France, nothing apparent would indicate that some of the best culinary products of the world are made here. On the hills to the east lies Château Margaux, which produces red wines that are sold for the price of modest cars. And here you will also find rows of oyster *cabanes.* Hardly any of them are particularly pretty. Mud and silt, along with the odd pile of empty oyster shells covered in flies, will keep any romantic feelings at bay. The area looks

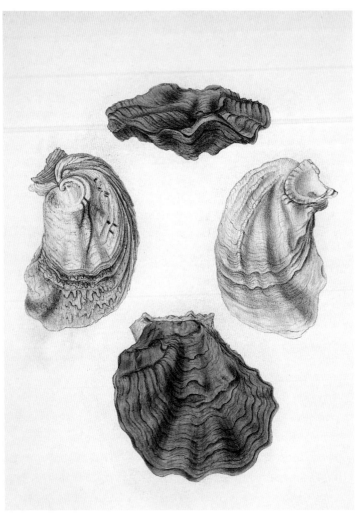

European flat oysters (top and bottom; dark and flat) in comparison to Pacific oysters (left and right; lighter and elongated).

more like a silty landscape devoid of life. However, if an oyster spends the last six weeks of its farm life here, in one of the seven hundred depuration operations, then it may receive the most noble title of all oysters—"Marennes-Oléron."

In Europe, a person can only very rarely find locally farmed indigenous oysters that were born at sea. The European flat oyster comprises only 1 percent of European oyster production. It is rounder and flatter than the Pacific oyster, which accounts for 99 percent of production in European oyster farms. This 1 percent of European flat oysters primarily originates in the estuary of the Belon River.

In Europe, until the middle of the nineteenth century, almost the only oyster found living on the seabed of the coastline was the European flat oyster, *Ostrea edulis*. It has been threatened with extinction for a long time; it could be compared to a lone wolf among millions of tame dogs. Belon oysters are really born in the Belon, where they spend their life, and they are an entirely distinct variety. They are rarer and more expensive than the common Pacific oyster. The creature inhabiting the shell is a little tougher, less meaty, and usually does not taste as intensely of the sea. In any case, it is the sole legitimate heir of the almost-extinct European flat oyster *Ostrea edulis*. It lost its role as the foremost European oyster in the year 1868. A ship by the name of *Le Morlaisien*, loaded with oysters, sought protection from a storm in the Gironde estuary near Bordeaux. The ship's captain believed his shipment had gone bad, so he tipped the boxes of oysters into the sea. According to legend, those oysters, farmed in Portugal, almost replaced the local European oysters completely on account of their greater resilience. In France they were given the false designation of "Portuguese" oysters, despite the species likely originating in Japan. Today, however, there are hardly any of those left either.

OVER ONE HUNDRED YEARS LATER, the oysters in France became practically extinct due to a viral epidemic in the late 1960s. This was not necessarily caused by pollution of the sea—which usually does not pose a major problem for oysters—rather, absurdly, by the antifouling paint on the bottom of the boats the oyster farmers used when tending the oyster beds. The dwindling oyster populations were replaced with Pacific oysters of the species *Magallana gigas*, which is particularly large and fast-growing, as indicated by its name—*gigas* meaning "giant." Since then, tales of oyster decimation have repeatedly made the rounds. In 2013, it was a strain of the herpes virus that killed 80 percent of the oyster population. Nevertheless, people have managed to care for these marine creatures and ensure their continued existence. For instance, after the Fukushima nuclear disaster in 2011, which rendered the oysters there inedible, Europeans exported the Japanese oyster species back to its country of origin.

For a long time, it was considered impossible to breed oysters in a laboratory. Oyster farming was done under more or less natural conditions. Not even nutritional supplements were added to the seawater. It was impossible to farm oysters out of their natural environment, as is done when farming chickens, salmon, or pigs. While oysters languish in their natural saltwater environment, their food floats to them on its own. An oyster does not cause any environmental stress whatsoever; quite the opposite, it even filters out organic ballast from the seawater. Oyster farming was akin to a heavenly mode of organic food production—until two Rutgers University researchers registered a patent in the United States in July 1997. The public has heard little about it since then, and oyster farmers do not like to dwell on it either. The patent in question is number 5,824,841, dated October 20, 1998, and it quite suddenly altered a species that had already existed for millions of years.

Since then, the oyster, formerly one of the most fertile creatures on Earth, has been damned to lead an infertile life, for the most part. Its existence is maintained solely via artificial means. An excerpt of the patent in question reads:

> Provided by this invention are novel tetraploid mollusks, including oysters, scallops, clams, mussels and abalone. Also, provided are a method for producing the tetraploid mollusks and a method for producing triploid mollusks by mating the novel tetraploid mollusks with diploid mollusks.

This describes a method by which a male oyster, conceived in a test tube with four copies of its genome, is crossed with a female oyster, which has two copies of its genome like every other creature on Earth. The results of this crossbreeding—after cell division (four plus two divided by two)—are hybrid "triploid" oysters with three sets of chromosomes, which are no longer capable of reproduction, thus revolutionizing the entire field of oyster farming.

I decided to finally visit an oyster farmer.

THE MAN
WITH THE
DARK GLASSES

"Enthusiasm, I like to compare,
To oysters, gentlemen—beware!"
JOHANN WOLFGANG VON GOETHE, "Fresh Egg, Good Egg"

THE ELDERLY GENTLEMAN across from me does not speak any language other than French, which I cannot understand. He is, however, wearing a large pair of dark sunglasses. Café de Turin, where he awaits me at a small marble bistro table, may still appear classy through his dark glasses; in any case, it has certainly seen better times. It has been a landmark in Nice, at the very same location, since 1908 and is a famous institution among oyster connoisseurs. Its latest renovation must have taken place a few decades ago. No matter, as we are here to discuss the product displayed on crushed ice in boxes beside the front door—oysters. Here, they exclusively sell their own oysters, rather than those from the market—oysters produced by

Café de Turin in Nice has been serving oysters from Marennes-Oléron for more than a century. Nowadays, the café is also frequented by women.

the Roumégous family. The family has had an oyster farm on the northwest coast of France in the Marennes-Oléron region since 1891.

When I ask Jean Roumégous—the name of the man with the sunglasses in Café de Turin—how long the oyster farm has been in his family, he counts all the fingers on one hand and several fingers on the other. All his figuring indicates that he himself already represents the fifth generation of oyster fishers and farmers, although his son François and his grandsons Adrien and Jules have already taken over the arduous work out at sea and in the offices.

Around a thousand tons of oysters are produced by the family enterprise each year. Jean oversees Café de Turin in the style of an elder maestro and resides here in a way fitting for a mature gentleman. With an almost inapproachable air, he sits in his café, heartily greeting the regulars. He is friendly to me as a guest yet maintains a noble distance. With the help of his business manager Alexandre, who interprets for us, we are able to hold a rather convoluted conversation. I ask whether there are still any piracy problems at the oyster beds. It is legendary that Jack London first enjoyed life as an oyster pirate and then as a guard—historical mysteries are still being written about this topic nowadays. Jean mildly shakes his head. Have they had any other troubles with their oyster production? No, since they have started buying triploid oysters, they have not had any variations in quality. As mentioned, mollusks with a triploid set of chromosomes are incapable of reproduction; however, they also grow much faster and are less susceptible to disease because they do not need to expend any energy on tiresome sexual reproduction. It is absurd that these oysters thrive on account of their infertility. The oyster breeder tells me that the triploid oysters were a blessing for the sector. He says this as though it were as inconsequential as a complete transition from black cattle to red heifers.

Triploid oysters do not require special labeling according to regulations, since they are not considered "genetically modified," as their set of chromosomes has only been reproduced but not changed. In the meantime, no one eating an oyster really knows what they are consuming. You cannot tell how many chromosomes an oyster has just by looking at it. Some people claim that artificial triploid oysters, introduced on the market by IFREMER—the French national institute for ocean science and technology—taste better than those born at sea like Aphrodite. In the Mediterranean oyster beds of the Étang de Thau, the breeders have even managed to farm triploid oysters that mature within nine months, instead of the usual three years. These can also be harvested year-round, since they no longer spend the summer months at the strenuous task of spawning for reproduction, like their amorous ancestors. The old rule of avoiding brooding female oysters during the months with an "R" has become less necessary than ever. In earlier times, it only served to protect the oyster populations during their reproduction period.

Over the past two decades, the laboratory-bred oyster has thus become the market leader of a large industry, quite without notice. On a global basis, more than half of all harvested oysters are presumably triploids. Recently, however, a quiet protest has arisen in France. After a documentary film called *L'huître triploïde, authentiquement artificielle* (The triploid oyster, authentically artificial) was screened, resistance formed against this method. A group of oyster breeders has come together by the name of Ostréiculteur Traditionnel (Traditional Oyster Farming). Their claim to fame is that their oysters were born at sea rather than in artificial tanks. Their battle call "born at sea" (*nées en mer*) promotes their natural condition, which could be taken for granted, as a rare sign of quality. However, these traditional oyster breeders are a minority, a tiny group of small artisan oyster farms. Others do the large-scale business.

French oyster farming associations, in addition to well-established oyster breeders, such as Monsieur Roumégous, attempt to pacify those who are worried. In the case of fruit, genome alteration has been taking place for years, including species of apples such as Jonagold or Belle de Boskoop. There has also been substantial progress with experiments to increase the genomes of salmon, trout, and goldfish. However, these crossbreeds have not been approved for the market, in contrast to the oyster. The greatest danger, which even the utopian genetic engineers of IFREMER admit, is that triploid oysters—a few of which do exist in nature, having "escaped" from the oyster farms—are mysteriously not 100 percent incapable of reproduction. Researchers do not actually know why, as it seems almost biologically impossible—or they may simply not want to give away the secret protecting their business model. Critics are worried that triploid oysters, hardier than their natural relatives, could quickly displace the latter if they are able to escape into a natural habitat. Then the triploids might also become extinct, due to their own reproduction problems—the oyster feast would be finished!

Back at Café de Turin in Nice, I have been hearing all about the demanding work of an oyster farmer. It requires fifty to seventy-five steps of work before an oyster can spend its final moment on the tongue of a connoisseur. Roumégous oysters are transported between Utah Beach in Normandy—where the family has owned several acres since the Allied forces landed there in World War II, since it is both historically significant and optimally located for oyster farming—and their home base in Marennes-Oléron. They grow better in the wilds of Normandy, whereas they acquire their famous taste during their final months at home in the calmer waters of Bourcefranc-le-Chapus. During the winter, the gills of these oysters attain their rare greenish color due to the unique blue algae that only exist here, and for which this family has been known throughout the world for generations. Those are said to be the best, and I will be able to try them soon. In fact, the rare *fines de claires vertes* were the first oysters to receive the French quality assurance of "Label Rouge" ("Red Label") for high-quality food products.

Despite all the French hospitality, and despite the touchy topic of triploid oysters, holding our conversation is not easy. Monsieur Roumégous is either too noble, too evasive, or he does not take me too seriously, since I am not a French-speaker and have refused to join him in a glass of wine before noontime. He does not really tell me anything that cannot be found out via research—until I recall an old journalistic trick. If an interview partner does not want to give away anything about themselves, it is best to ask them about their neighbors. Without harboring any lofty expectations, I ask him what he thinks about the market leader among high-quality oysters, the famous company Gillardeau, with its production site right next to his? Gillardeau oysters are considered the Rolls-Royce of oysters in gastronomy and fish markets—an expensive, high-end product.

Suddenly the elderly gentleman and his business manager, who is interpreting for us, become livelier. Their eyes begin sparkling, and they suddenly both speak to me at the same time. The Gillardeau oysters, for sale at elevated prices everywhere in delicatessens, are not actually from the Gillardeau family, scolds the oyster farmer. They have been bought all over the place and are merely packed in their factory. It could constitute fraudulent labeling. If you look at a box of Gillardeau oysters, you will see mollusks of various origins with shells of assorted sizes and shapes. They come from everywhere, while Roumégous oysters, his own oysters, are raised on his own farm, cared for and turned by hand. To demonstrate this, Alexandre, the business manager, wildly flips his cell phone back and forth on the marble bistro table. That is how they care for their oysters, while the neighboring company, Gillardeau—more famous and considered the archetype of oysters—just buys their oysters elsewhere and sells them for unreasonable prices. They are a wild mixture, a blend. His Gillardeau neighbors are simply good businesspeople, not oyster breeders. In the Roumégous baskets outside Café de Turin, you can see that all his own oysters look the same.

Paying a visit to a good fishmonger, who naturally carries Gillardeau oysters, confirms what I have been told. Those oysters have been raised on one of the company's many oyster farms in France, Ireland, Scotland, or Portugal. This wide assortment of oysters from different seas then undergoes depuration during their final weeks in seawater tanks in Marennes-Oléron to receive the coveted quality label. Nobody at the Gillardeau company wants to talk to me about this. When I politely request

A whole company of women coming ashore with the tide:
Auguste Feyen-Perrin, Return of the Oyster Fishers, 1908.
Nowadays, men also work on oyster farms.

an appointment with them during my oyster research trip to Marennes-Oléron, I do not receive any response whatsoever. Astonished, I stand in front of the company's noble industrial hall, recently built just shoreside of Île d'Oléron. Its sterile gray shape looks so vastly different from all the *cabanes* located on the smelly canals where the artisans farm their oysters. I bought a few Gillardeau oysters from an automatic oyster dispenser located in front of their company headquarters.

The Gillardeau company is practically as old as its smaller competitor Roumégous. It was founded in the year 1898 by Henri Gillardeau in La Rochelle. The tale goes that Henri could not read, although he could count. After making a fortune during the height of the first oyster boom, he had a motto affixed to his new house—naturally right across from the mayor's house—*Ca m'suffit.* ("That's good enough for me.") His grandson, Gérard Gillardeau, flipped the family slogan upside down at the end of the 1970s and turned the nameless, every-day product into a brand that carried his name and became the market leader around the world—or at least around Europe. The higher quality of the Gillardeau oysters was explained by the fact that fewer oysters were contained in the oyster bags, which encouraged the growth of the shellfish competing for plankton. Gillardeau produces around two thousand tons of oysters annually. Their greatest marketing gimmick has been to laser a "G" onto their oyster shells since 2004, to purportedly prevent product piracy. However, if you buy a box of Gillardeau oysters, the rather pompous laser brand is the only thing those oysters will have in common—that is, if you believe the slightly less successful neighboring competitor, Roumégous.

With a "G" on the shell, the Gillardeau oysters have been branded, like a print by Albrecht Dürer (AD) or a Rolls-Royce (RR), unmistakably positioning them on the market. The primeval shellfish has arrived at the international market of brands. Very few oyster farmers can compete. The company Maison Tar-bouriech, with its pink-tinted oysters farmed on ropes in the Mediterranean cove of Étang de Thau, is one of them. They have developed their own luxurious oyster spa complete with baths and salts at their headquarters—where Casanova is also said to have spent the night. Other oyster enterprises have taken different approaches. In 2004, a comparably young family enterprise by the name of Parcs Saint-Kerber had the idea of naming their

During the Oyster Wars in the 1880s, oyster pirates illegally harvested oysters in Chesapeake Bay by dredging metal baskets along the seabed, leading to violent skirmishes.

French oysters after the Russian "Tsarskaya." The attempt to elevate them to the level of a luxury product has had some success in the delicatessens of Central Europe, even though no legend confirms that a tsar ever ate any of their oysters.

The only commercial oyster farm in Germany, located on the island of Sylt, is just a little older. Oysters had already become extinct in the tidal flats of the German coastline by the 1930s at the latest. Dittmeyer's Oyster Company has successfully marketed their product by the name of Sylter Royal since 1987. Of course, no "royal" has ever eaten any of these oysters, and they were not born on the island of Sylt either. The spat are purchased from Swedish or Irish tanks. In winter, the shellfish must be removed from the chilly water of the North Sea and transferred to specially built, 885-foot-long heated tanks, because the delicate creatures would not otherwise survive the cold winters in northern Germany. At an output of only one million oysters a year, or less than one hundred tons, the annual production in Germany is extremely low.

There are hardly any traditional oyster fishers left anywhere in our day and age. Only in North America, primarily in the Gulf of Mexico and in Chesapeake Bay, is traditional oyster fishing still practiced in a few places by harvesting the shellfish from the seabed using dragnets. In the meantime, 96 percent of the global production comes from oyster farming in aquaculture. Worldwide, there are four different methods for cultivating oysters.

In conventional oyster farming on the seabed, oysters are seeded on the bed of the sea and harvested after several years by dredging with nets. Naturally, this first method, called bottom or beach cultivation, still used in a few places in England, results in significant losses. In France, the second method, so-called rack and bag cultivation, is most widespread. In the Atlantic tidal zones, the flood tide provides the oysters with daily nutrition, while the oysters are tended to with heavy machinery during low tide. The oysters are kept in polyethylene nets, which are held on iron racks around one-and-a-half feet above the seabed and are turned and tapped several times during low tide. This method protects the oysters from their natural predators and keeps away the mud that would otherwise inundate them.

The third and most technically complicated method is "rope cultivation," which involves gluing juvenile oysters with cement to ropes, or keeping them in small nets—work usually done by hand. Particularly in the Mediterranean, where the difference between low and high tide is insufficient, they are regularly dipped into the water and lifted out again using timers. This method has once again enabled oyster farming in Montenegro, in Croatia at the estuary of the Lim Bay, and even just recently in the Po River Delta in Italy. Raft cultivation, the fourth method, is similar; oysters are hung from bags on a raft in the sea. In China, where 80 percent of the world's oyster production takes place, oysters are hung from colored buoys, enabling the oyster farmers to tell the size of their shellfish according to the color of the buoys. In contrast to Europe, where oysters are farmed for the luxury market, they are produced in China as a source of protein for the general population, as well as for making oyster sauces and smoked oysters in tins. Chinese oysters are not a luxury product at all; they are more like industrially farmed produce, such as soy or corn.

Toward the end of my visit with Jean Roumégous, I am taken to his stands outdoors, where oysters are on display for passersby. He proudly shows me how each oyster looks just like the others. Much like his competitor Gillardeau, he also carries more expensive *spéciales de claires*, in addition to *fines de claires*, the former of which have undergone depuration in tanks for at least two months longer. Silently signaling a waiter to shuck a dozen oysters for me, he shows me the unique green gills of these creatures. Gillardeau oysters do not have those. Then Jean Roumégous bids me *au revoir* and trots back into his café, where he enjoys another little glass of wine while awaiting his regulars. His oysters were very tasty.

Carlo Ponti, Oyster Vendor in Venice, circa 1850.

ONCE AROUND
THE WORLD

"Why, then the world's mine oyster,
Which I with sword will open."
WILLIAM SHAKESPEARE, *The Merry Wives of Windsor*

MY JOURNEY AROUND THE WORLD begins at the shores of Normandy. The Atlantic Ocean, where oysters grow, and which I must cross, has just retreated from the harbor of Trouville-sur-Mer. Almost nowhere in the world are the tides as high as this region, which has become a center of oyster cultivation. The oyster farmers can drive their tractors into the widespread oyster beds to tend the shellfish. They are kept in plastic bags on racks, where they are regularly turned. When the tide comes flooding back, the tractors return to land, and the sea provides a free smorgasbord of nutrients. In the evening, the boats in the nearby harbor are stranded in the mud. Here in the oysters' homeland, we are far from flagship stores and marketing gimmicks. The seafood is not labeled according to brand name at the traditional *marché aux poissons* (fish market), but according

The journey begins in Brittany: Eugène Boudin, Le marché aux poissons, *1875.*

to origin and size. There are six sizes of oyster, from no. 5 (which weighs just under two ounces), to no. 0 (which weighs over five ounces). The most popular is no. 3, weighing a light two-and-a-half ounces.

The origin of the oysters sold here—often eaten on the spot—lies only a few miles away. Nowhere can you more clearly recognize the influence of the seawater and the specific region on the taste of the oysters, sold by the nine fishmongers here with names such as Côté Mer or Chez Pascal.

I immediately notice the Huîtres Spéciales d'Utah Beach n°3. Did the Allied troops really pass over these oysters when they landed there to liberate the world from Nazi Germany?

Right next to them in a wooden crate are Saint-Vaast-La-Hougue n°4 from several miles farther north. Of a much closer origin are those from Asnelles & Isigny n°2. A few baskets further on, I see those from Pleine Mer n°3. The fishmonger is happy to put together a selection for me. I make a real effort to note the differences and log my remarks in my digital notebook. When tasting oysters, we differentiate three to four distinctive characteristics. First, we note the saltiness, which varies according to the sea where they originated, followed by the texture. And finally, we may note a "fruity" taste before we chase the oyster down with a cold beverage. Or, as Ernest Hemingway more fittingly put it in his book *A Moveable Feast*:

> As I ate the oysters with their strong taste of the sea and their faint metallic taste that the cold white wine washed away, leaving only the sea taste and the succulent texture, and as I drank their cold liquid from each shell and washed it down with the crisp taste of the wine, I lost the empty feeling and began to be happy and to make plans.

I had planned my journey to follow the taste of the oysters. In fact, the four kinds of oysters I tried on the beach in Normandy all tasted a bit different. The historically laden ones from Utah Beach had slightly darker, almost black cilia. Those from Asnelles—I see now with astonishment in my notes—were rather tough in comparison. The so-called Pleine Mer oysters were closest to whatever dish I had eaten at the stand in the fish market—although I no longer recollect what that was.

Pleine Mer oysters are farmed farther away from the coast in Brittany or Normandy. Due to the significant difference to the ebb, the flood tide still washes fresh seawater over them daily, whereas they close during low tide. Theoretically, they should taste slightly more of iodine and seawater than the *fines de claires* from the oyster beds closer to shore. I have noted that the legendary Utah Beach oysters, which are counted among the best in the world with their firm and plump saltiness, also have a fruity taste. Ever since the French president François Mitterrand, known as a gourmet, praised the oysters bred by oyster pioneer Georges Quetier, Utah Beach has been considered a noble source of oysters. However, the oysters are only kept there during the final months of their lives, on the windy and plankton-rich side of the Cotentin Peninsula. Those final months in the stormy sea are said to make them particularly strong and pure. For young oysters, the surf would otherwise be too heavy, and the location too costly for the oyster farmers.

ON THE OTHER SIDE of the English Channel, oyster farming is also venerated. I continue my journey to Whitstable, in the estuary of the River Thames. The Romans already exported oysters from here to Italy; there are reports from the year 78 C.E. that oysters were carried over the Alps by apes in cooled boxes. Each year an oyster festival takes place here, including an oyster eating competition. The town was founded because of oyster farming.

At the stands of the local oyster farmers, so-called wild oysters are available in addition to cultured oysters. The wild

Cultural meeting place, as well as social event: The First Day of Oysters, *1861.*

oysters can attain huge sizes and are collected by oyster fishers on the sandbanks, just like in the old days. I buy the largest
oyster of my life at the beach there, a "wild" oyster, which I
have trouble swallowing. According to the oysterman, it is
around nine years old. The body of the oyster alone is about as

big as the palm of my hand. It's almost impossible to get the sandwich-sized creature into my mouth in one piece and swallow it. I have to admit—I gag on it more than I enjoy it. In addition, it seems a little tough. It's more for the purpose of research than pleasure.

Nevertheless, right nearby I make a discovery of nicely cooked oysters. "Most oyster cookery is misguided," my travel guide Rowan Jacobsen advises in his book *A Geography of Oysters.* "Why take a delicate, fleeting flavor and disrupt it with heat?" Yet here in the nearby The Sportsman—a highly praised, yet traditional gourmet restaurant—I dine on three different, well-prepared oyster dishes. The cook there, Stephen Harris, dares to add something of his own to the local oysters, even though these creatures were his nightmare at the beginning of his career as a cook. Before Harris opened his pub-like restaurant in Seasalter, which was soon considered one of the best in England, the self-taught chef had a repeated nightmare about a pile of broken knives and broken oyster shells. As he says, "How could I hope to open a restaurant when I couldn't open an oyster?" His angst could explain his rather aggressive approach when faced with preparing them. In my opinion, his poached oysters in *beurre blanc*, with pickle and avruga caviar, might be the only dish that comes close to tasting as delicate as a fresh oyster. Harris is happy to tell me his recipe. He makes a white sauce with a pickle, a shallot, a little white wine, cider vinegar, butter, and cream. Then he poaches the oysters in their own water and adds a little avruga caviar. It sounds wonderful; however, avruga caviar is just a cheap ersatz made of 40 percent herring, in addition to lemon juice and cornstarch. The dish is a tasty protest against the *haute cuisine* that I intend to enjoy later.

I naturally did some research prior to my worldwide search for the perfect oyster. The leading restaurant critic in Germany, Jürgen Dollase, reports he once heard from a fish specialist in Brittany that the taste of oysters develops best at human body temperature. This led Dollase to conduct a sensory experiment, as he called it, which was unsuccessful. He carefully heated his oysters to body temperature before eating them. "At this temperature, the oysters developed a very strange, almost unpleasantly strong, penetrating aroma, so that the meal was not enjoyable," he reports. The chilled freshness of oysters ensures their good taste, just like the drinks that accompany them, such as champagne, white wine, or beer. Only after the oyster has been slowly warmed in the mouth and chewed do the aromatic notes develop, which might otherwise be perceived as negative. Dollase sums up that "if the oyster is served warm, it lacks the combination of aromas perceived when it is slowly warmed, yielding a gradual blend of many aromatic components."

Because oysters taste best chilled, *haute cuisine*—which always strives to prove its genius via cooking techniques—tends to completely avoid them. Oysters are served in good restaurants for steep prices, but they hardly play a role in the menus of *haute cuisine* with its star *chefs de cuisine*. The reason is that each addition—apart from a little lemon juice; or the classic *sauce mignonette* of shallots, vinegar, and pepper (which is refused by most oyster connoisseurs)—tends to reduce the taste of

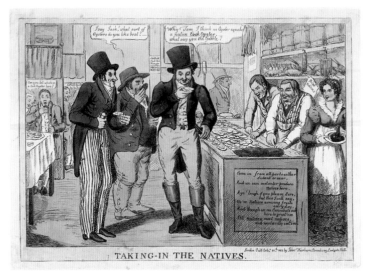

TAKING-IN THE NATIVES.

Edward Dando, known as "the Oyster Eater," was frequently imprisoned because he often ate dozens of oysters at the bar without being able to pay for them. He died in prison in 1832— after catching cholera there.

the oyster rather than refine it. In its raw state—which already includes packaging, plate, and spoon in the form of the shell— the oyster is more like fast food than a composition prepared by gourmet chefs. Nevertheless, there used to be cookbooks made up exclusively of oyster recipes. Today, they seem rather bizarre; recipes that are entirely unthinkable nowadays can be found in them. For example, the author of *The Three Musketeers*, Alexandre Dumas, considered his *Grand dictionnaire de cuisine* (Great Dictionary of Cuisine) his major literary work. In it he suggests frying cod, which he did not particularly like, along with six dozen blanched oysters to make the former palatable. On the other hand, in the twentieth century, Rudolf Kilias reports in the addendum of his book *Austern* (Oysters) that oysters can be made into sausages. This involves grinding the meat of a leg of lamb, seasoning it, adding fifty oysters, some baguette bread, four egg yolks, and a slice of lemon. If one does not want to cook the sausages immediately, Kilias notes that one can cover them up, and "they will keep for a while." Trying this out is not recommended.

Nevertheless, I was curious what perhaps the world's most famous cook, René Redzepi—who became famous for promoting pure, highly local ingredients in his restaurant Noma in Copenhagen—would do with oysters, which naturally originate on the Danish coastline. Redzepi preaches culinary unity with nature. He has uploaded the recipe for his signature dish, "steamed oyster," as a short film to the internet. For it, he collects several herbs, which he calls "beach onions," from the seacoast, where his oysters originate. He then seasons the oysters on the half shell with horseradish and capers from elderberries and decorates them with blossoms from the sea. Next, after replacing the top half shell in a small, covered pot with seawater and stones collected on the coastline, he steams the oysters for four minutes, then immediately serves the pot to his guests.

That was the concept. But once I had finally gotten a reservation at Noma on my search for the perfect oyster, Redzepi had long forgotten this recipe and given up serving it, reverting to the original form of the oyster. He served the oysters raw, just as nature made them, only seasoned with a little pepper, herbs, and a flower. And to leave a trace of the genius *chef de cuisine*, he served the juice of a half-ripe Danish tomato, instead of the commonplace juice of an Italian lemon. It represented the capitulation of a star chef in the face of the ingredient. And it was very tasty.

The journey continues from devoutly Lutheran Copenhagen to sensually Catholic Italy. In Modena, an Italian by the name of Massimo Bottura occasionally challenges René Redzepi for the title of best cook in the world. When Bottura serves an oyster in his restaurant, then it is not an oyster, but just a reminder of one. It is a memory of the Oyster Conversion Experience (OCE), a recollection of the first oyster he ate in Normandy. Bottura's signature dish, "Normandy," tries to create the perfect image of an oyster without using the flesh of an oyster. For it, Bottura serves some raw ground lamb, minced and colored with salt and pressed peppermint. It is bedded in an original oyster shell, lying in a cream of oyster water and cider sorbet, and decorated with seaweed, simulating the darker innards of an oyster. This dish is more closely related to cooking philosophy or an edible history of humanity than *haute cuisine*, although it tastes fantastically good. Bottura connects us with the original experience of humankind. He knows exactly when he consciously ate an oyster for the first time. "I'd never really tasted an oyster until I tasted one in Normandy," he admits in his autobiographical cookbook *Never Trust a Skinny Italian Chef.* However, he does not consider his recollection personal; rather it is the collective experience of humankind. "My memory of Normandy began well before reaching its shores; it was tangled in the collective references to history…"

Bottura has attempted to recreate the taste of an oyster with ingredients that are foreign to shellfish, so that he does not have to serve his guests the cannibalistic, slippery rawness of an oyster. Bottura's Normandy is the apotheosis of culinary art, a process of emancipation, humanity's self-assured farewell to the Dark Ages of eating live creatures. Instead of a barbaric act, during which a highly developed creature feasts on a simpler one, swallowing it in a single bite, we are confronted with a complicated process of sensory mimicry. Bottura is attempting to enjoy an oyster without having to consume it. His dish is philosophy. Bottura is an artist and the Pygmalion of oyster chefs. A cook challenges the world to create the perfect dish anew. Like a god, the best cook in the world wants to prove that they can compete with nature, that they can create the perfect dish, as though it were not already lying on the seabed. "In our 2011 recipe we wanted to invent the world without inventing anything at all," Bottura writes about his dish. He reconstructed the first dish of early humans, at the end of humanity as we have known it to this day—"without being an oyster," Bottura declares proudly with an impish laugh, as he serves it to me.

The difference between a raw oyster and Bottura's sophistic non-oyster dish is more than a culinary joke. It corresponds to the difference between a "cold" society and a "hot" society, as defined by Claude Lévi-Strauss, the famed anthropologist, in his 1964 philosophical work *The Raw and the Cooked*. "Cold" cultures invent

complicated structures to avoid having to change themselves—
they prefer eating oysters raw—while "hot" cultures are constantly
reinventing themselves. Massimo Bottura clearly represents
the latter faction. To observe a "cold" culture enjoying oysters, it
is worthwhile leaving Italy and crossing the Atlantic to North
America, where the world of oysters is different.

Yet first, we'll take a short detour to the fishers of Apulia,
where we can learn something about the dialectics between
domestic and foreign tastes. Here in southern Italy, wild oysters
are still gathered by oyster divers. There are also a few new oyster
farmers who have started operating here, sending their rather
expensive oysters to the restaurants in Rome, just like in ancient
Roman times. In in the harbor of Bari, there are hawkers sell-
ing fresh shellfish, tasty sea urchins, and—sometimes—oysters
on old tin boxes. They are small, unattractive-looking, and have
a strong taste. They have clearly not been cleaned or purified
and—although they are likely fresher than almost anywhere
else—I have to steel myself somewhat to eat them. When I ask
the proud fishmonger with dirty hands where his oysters were
from, he falsely claims they are from France, and particularly
good. To him, a local Italian, the oysters he collected himself,
which he had just sold me, were of lesser value.

A little farther south, in the tourist-filled harbor of the town
of Gallipoli, down at the heel of Italy's boot, there are more
noble fishmongers, who sell vacationers fresh shellfish with a
glass of white wine. If you order half a dozen oysters from the
salesperson (experienced with tourists), you will receive beau-
tifully shaped and purified model oysters. If you praise them

and ask where they are from, the salesperson will tell you, just as falsely, that they are local oysters from Apulia, which grow in the sea here. A practiced eye might note the little white label on the yellow plastic net—required in the European Union—which documents the packaging date and the French origin of these oysters. I have rarely experienced the differences between domestic and foreign tastes as clearly as here. At both fish markets I traveled to, in Bari and Gallipoli, the fishmongers were dishonest to me. The professional salesperson knew that tourists expected something authentic yet sold them an industrial product instead. And the amateur in Bari played down the value of the oysters he collected himself as supposedly lesser quality, claiming they were a noble industrial product instead. Tourists want something special and are sold an everyday product, while the local population either prefers the purportedly better foreign product, or they enjoy the local oyster specialty—"red oysters." Here in the southern Mediterranean lives a dark red relative of the oyster, the rare thorny oyster (*Spondylus gaederopus*), which is collected individually during arduous work by oyster divers from a depth of at least thirty feet. The dark red color of their flesh comes from the sponges that attach themselves to their spiny shell.

THERE IS NO LACK OF different oyster varieties on the other side of the Atlantic, in North America, the land of the wild and free oyster species, a continent that owes quite a lot to oysters. The oldest oyster bar in the United States is considered the Union Oyster House in Boston, founded in 1826—making it so old that

Modern shell mounds: two New York oyster bars under Manhattan Bridge, photographed by Berenice Abbott on April 1, 1937.

it has been incorporated into the list of protected landmarks in America. Slightly younger, but just as venerable, is the oyster bar in Grand Central Terminal, opened in 1913 in the vaulted hall of the central train station of New York City—an area where the Indigenous Lenape people had harvested oysters and other shellfish for millennia. In the nineteenth century, one million oysters were consumed daily in the city that never sleeps. In the noble Grand Central Oyster Bar, various oyster species from the east and west coasts of North America are served on red-checked tablecloths, just as they were decades ago.

Your order of OYSTERS A LA ROCKEFELLER Since 1889 when this dish was concocted
by Jules Alciatore at The Restaurant ANTOINE is number: _____ 1262315 _____

| The Restaurant ANTOINE | | 713 St. Louis St. |
| Founded in 1840 | | City of New Orleans |

Present Proprietor Roy L. Alciatore, Son of Jules and Grandson of Antoine Alciatore,
is sampling the millionth order. *The recipe is a sacred family secret.*

COPYRIGHT 1938, BY ROY LOUIS ALCIATORE

*A culinary dish as a work of art: a numbered, but unsigned
receipt for "Oysters a la Rockefeller" at Antoine's restaurant
in New Orleans.*

In North America, you can find opulent coffee-table books
depicting the various kinds of oysters, quite comparable to
varieties of wines. Some of the oysters available are Kumamoto,
Nonesuch, Fat Bastard, Resignation Reef, Baywater Sweet, Hama
Hama, and over three hundred others. Not only do they look
different, but they also taste different according to the marine
territory—or merroir—where they originated, just like different
wines.

Here at the Grand Central Oyster Bar, oysters are also prepared in various ways, whether raw, fried, steamed, or as Oysters
Rockefeller—a classic oyster recipe, guarded with the same
secrecy as that of Coca-Cola or Averna. To enjoy the original
Oysters Rockefeller, we need to continue our journey to New

The oldest oyster farm in California, the Tomales Bay Oyster Company, has been operating since 1909 from its location north of San Francisco.

Orleans. No one knows the secret recipe other than the *chef de cuisine* at Antoine's, where the recipe was invented by Jules Alciatore in 1889.

Legend has it that Alciatore created the dish of baked oysters with the spices he happened to have on his counter, naming it after the richest man on the planet at the time, John D. Rockefeller, as the dish tasted so incredibly rich. The original recipe has never left the kitchen of this restaurant; it is one of the best-kept culinary secrets on the planet. The owner of Antoine's claims that all recipes for this dish found in cookbooks are fakes.

Each time the dish is served at Antoine's, it is numbered like a work of art, and the guest is given a receipt.

The oyster journey across North America from the Atlantic to the Pacific is—according to Rowan Jacobsen's guide *A Geography of Oysters*, which I trust—like the journey from a Riesling (simple, but with an incomparable mineral taste, if well grown) to a Sauvignon blanc (fruitier, more aromatic, and not easily classifiable according to taste). After stopping in Tomales Bay, north of San Francisco, where there is nothing other than two oyster stands—not even a spot where you can eat them—my trip ends in one of the greatest palaces of oyster tasting. At 1517 Polk Street in San Francisco, there is a little bar with the unimposing name of Swan Oyster Depot. Since 1912, it has been ready for business during its short midday hours, and it is nothing more than a long marble counter with exactly eighteen barstools, occupied by exactly eighteen guests, who have ordered plates of oysters and other seafood. The Swan Oyster Depot is the best place in the world to eat oysters. It is a restaurant reduced to the form of a skeleton, and because you cannot make a reservation, a line forms in front of the door up through Nob Hill long before the restaurant opens each morning. While we wait for over an hour to take a seat on one of the barstools, we are served one or two beers. And we have plenty of time to look through the display window—which has the antique, curved lettering of a saloon—where we can see the fresh seafood that is served inside.

Inside, there is not a menu to be found. The dishes have been handwritten on three panels above the bar and have not changed over the past thirty-five years. Everything in this bar

Unchanged since 1912. The original owners at the Swan Oyster Depot in San Francisco in 1934.

is reduced to the essentials. The available dishes are few, considering that we are in one of the most revered oyster bars in the world, a place with cult status. Only three or four species are offered here: usually the notorious Blue Point (Long Island), Miyagi (Pacific Northwest), and the little Kumamoto (California). Nothing else. The attentive bartender knows the sequence you should follow when eating the oysters and arranges them accordingly on the plate. Nothing is exquisite here in the Swan Oyster Depot. The oysters are quite self-sufficient. There are no special ingredients, nor any special oysters; the plates are a bit chipped, and there is real enthusiasm surrounding the authenticity of this unique bar. A handwritten sign hanging from the ceilings warns: "Attention! Swan Oyster Depot does NOT have a WEBSITE!! Anything you see online is unauthorized!! We only deal person to person!! NO WEBSITE!!" This is the oyster turned into a bar—rather inaccessible, but very tasty. Nothing is better or more correctly prepared than anywhere else, if you want to speak of cuisine. Everything is natural and pure. If I were an oyster, I would like to be served here.

If I crossed the Pacific from here to Japan, then I would likely despair, considering my European oyster tastes. In Japan, oysters are often quite large and are breaded and fried, served with knife and fork and cabbage salad in a dish known as *kaki furai*. This is a dish that has been liberated from its nautical origins. On the other hand, traveling to the Caribbean beaches of Cartagena in Colombia, I found fresh oysters sold right on the beach—at over 104 degrees Fahrenheit in the shade—without any crushed ice, protected by nothing but a plastic crate from the heat, and only partially cleaned (but in all sizes). An experience, without doubt, yet limited in terms of a recommendation.

If all that sounds too strange, or if Colombia is too far away, and eating fresh-gathered oysters on the beach seems too dangerous, then a person could resort to global delivery of the creatures in the form of tablets. A French start-up called P.O.P (Pure Oyster Flesh Power) packs pulverized oyster into capsules, which deliver the positive effects of eating oysters to those people who refuse to eat live creatures. According to the unsubstantiated marketing claims on the packaging, the capsules of oyster powder are said to fortify the immune system, help strengthen the bones, protect the nervous system, and guard against tiredness and hair loss, among other alleged benefits.

A hand-colored advertisement printed for a London oyster seller in the early 1800s.

FRAGMENTS ON THE WORLD HISTORY OF THE OYSTER

> "I sing the Oyster! (Virgin theme!)
> King of Molluscules! Ancient of the stream!
> Thy birth was Time's."
>
> JAMES WATSON GERARD,
> *Ostrea; Or, the Loves of the Oysters*

TO WHOM DOES THE WORLD BELONG? There are arguments that claim the speechless, immobile oyster—rather than the intelligent, hypermobile *Homo sapiens*—might rule the world. From the perspective of world history, human beings, with our capacity to read and write, are just an evolutionary off-shoot, a comparatively new species lacking a hundred million years of tradition. We can draft interesting theses about oysters, but they have more ancient rights to this Earth. Long before

Homo sapiens made an appearance, there were dinosaurs, and long before the latter died out, oysters had already come into existence. These marine creatures have survived the devastating impact of comets, ice ages, and every episode of global warming since. They served our prehistoric ancestors as food, without doubt. However, they almost have not survived due to humankind's insatiable hunger. In the event that human beings become extinct in some future century—or the living conditions on planet Earth make it uninhabitable—then the chances are good that the oyster, which already survived humankind's original sins, will also survive our own self-destruction by environmental degradation.

The dates marking the world history of the oyster are truly incredible. During the Triassic period, an unimaginably long time ago in the history of our Earth, beginning about 251.9 million years ago, oysters already existed all over the world. They are classified under the scientific genus *Gryphaea*, and their fossils are still found in fields and high on mountains around the world, even in the Alps. Oysters are even older than the mountains; during the Triassic period in prehistoric times, the Earth's landmass still formed a supercontinent by the name of Pangaea. Only later did the different oceans, mountains, and continents come into being. Oysters have experienced all of that, even if they could not see it with their own eyes.

The latest research, published in the wonderful *Journal of Molluscan Studies*, indicates that oysters evolved from the primeval soup shortly after the great extinction at the end of the Permian period, around 250 million years ago. In contrast to

Primeval oyster fossils, like the Umbrostrea emamii *pictured above, have been found around the world.*

many of their contemporaries, such as the early *Lystrosaurus*—similar to a pig with swordlike teeth—I read with great surprise that oysters also survived the next great extinction during the transition to the Jurassic period around 200 million years ago, when most of the existing forms of life on Earth died out once more. The next great catastrophe hit our planet at the end of the Cretaceous period about 66 million years ago. An asteroid with a diameter of nine miles hit the Yucatán Peninsula. The gigantic

George Frederic Watts, Tasting the First Oyster, *circa 1883.*

crater created a huge earthquake, tsunamis, an extended period of darkness, and an ice age, causing about 75 percent of all species on Earth to become extinct. I am astonished that oysters even survived this apocalypse undeterred. Oysters, such seemingly sensitive creatures, are apparently one of the most resilient forms of life on the planet Earth.

Admittedly, no one can really imagine how long 251.9 million years ago was. However, I had a wake-up call that showed me how literally commonplace oysters were. The epiphany happened one beautiful summer day in England. A turn of fate took me to the university town of Cambridge, where I became exhausted while visiting the venerable Sedgwick Museum of Earth Sciences. I had gone there to see the Beagle Collection, an assortment of stones and fossils that Charles Darwin brought back from the expedition on the HMS *Beagle*. However, only a single small display case was dedicated to that collection. A little disappointed, I continued walking without interest through endless rows of display cases containing thousands of fossils labeled with tiny, handwritten, yellowed cards exhibited in cardboard boxes. When I discovered a fossilized oyster (*Ostrea*) that was found in the chalkstone of Norwich, I was as excited as a young researcher. But to my amazement, I found dozens of fossilized oysters in the display cases going ever further back in the history of the Earth. Among the strange relics of extinct trilobites and primeval sponges, oysters were the only form of life that had survived as though no cosmic catastrophes had taken place since they came into existence.

To this day, these so-called devil's toenails can be found in fields everywhere in the world. They are primeval oysters with a highly curved bottom half shell, and they were so widespread in the Jurassic period 200 million years ago that they were chosen as scientific index fossils.

Many, many millions of years later, when humankind became dominant on the planet Earth around twelve thousand years ago, oysters had already long been residents here. Human beings might not have become what we are without oysters. We have been eating oysters from our early days. Wherever our earliest ancestors lived as hunters and gatherers, they also ate oysters. It almost seems as though the moment they settled down at a permanent site is connected to the consumption of shellfish. All early cultures have been found at sites where oysters were growing. The oldest traces of human life that archaeologists have found on Earth are not primitive tools or buildings, but shell middens, which are literal piles of debris from meals. In fact, they are prehistoric garbage dumps consisting primarily of huge piles of shells. They can be found in Scandinavia, as well as Mexico, along the Bering Strait, and in California. On the Hudson River near New York City, there are huge shell middens that are ten thousand years old and are regularly unearthed during excavations for building developments. The oldest evidence for the mass consumption of oysters was found at Pinnacle Point on Mossel Bay in South Africa. In 1997, a shell midden was discovered during excavations for a golf course, where tools and the remnants of paints were also found, indicating the first evidence of early humans. These were dated to 164,000 B.C.E. by archeologists.

Since 2007, Pinnacle Point has become one of the largest prehistoric archaeological excavation sites anywhere. There are scientists who claim that *Homo sapiens* could have spread around the world from this oyster-eating site, where evidence of fire has also been found. Instead of oyster knives, fire hypothetically served to open the oysters, just like today in René Redzepi's restaurant. Collecting oysters was easier than hunting, and unlike berries, oysters were available year-round.

Oysters contain plenty of protein, few carbohydrates, important minerals, and lots of vitamins, which could also have sufficiently nourished early humans without the need to hunt, collect berries, or cultivate fields. Yet to my knowledge, no one in our day and age has assessed whether a person can survive on oysters alone.

Excavation of a shell midden on Elizabeth Island in the Strait of Magellan, 1888.

It might seem that human beings have not learned a lot since then. Shell middens provide evidence that prehistoric people did not always lead a sustainable way of life. At the bottom of the shell mounds are usually the largest shells, while the remaining shells higher up are smaller and smaller. This continued until the oysters, which grew slower than the human appetite, were all eaten up and became extinct in certain places. It was not much different in the Holocene than ten thousand years later with the French and the Americans, who decimated the oyster populations around the world in the nineteenth century, when they were sold on the streets as food for the masses.

It is certain that oysters have survived the greatest crises of our planet for more than 250 million years, including meteorites, ice ages, and the submerging of entire continents. On the other hand, it is not at all certain whether *Homo sapiens*—at only a few hundred thousand years old—will survive the next few centuries. In fact, upon visiting the museum in Cambridge, it became clear to me that we live on a planet of oysters. In terms of the history of the Earth, the genome of the oyster has survived for much longer than that of humankind. And according to the theory of the "selfish gene" proposed by British biologist Richard Dawkins, the actual evolutionary subjects are not the individuals, but their genome. Their genes want to survive, so they form living creatures, biological engines, to ensure their survival. Their propagation only serves to ensure the survival of the genome. Individuals can die; however, according to Dawkins's thesis, they only exist so they can reproduce the information stored in their DNA as often as possible. To survive, DNA has found various carriers—intelligent creatures, such as human beings, or simpler ones, such as oysters. From this perspective, humankind does not epitomize the crowning feat of nature; rather, it is nothing but a passing appearance that "must be overcome" in evolutionary terms according to Nietzsche's philosophy—possibly by the oyster, which is only temporarily providing nutrition for human beings on this planet.

Human beings could really learn something from coexisting with the oyster. The 1877 book written by German zoologist Karl August Möbius, *Die Auster und die Austernwirtschaft* (The oyster and oyster farming), is thought to be one of the first books written from a holistic ecological perspective. Considering the interaction between human beings and oysters at oyster

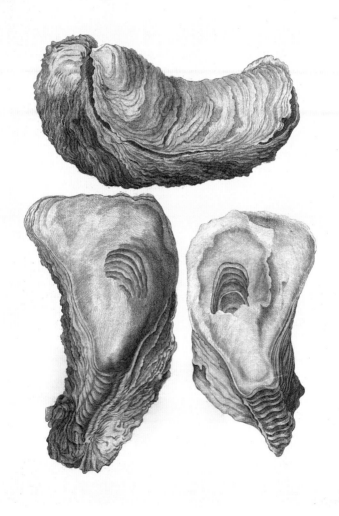

Primeval creatures of the future: over the course of natural history, oysters, which have existed much longer than humankind, might also outlive us.

beds, Möbius coined the term "biocenosis," meaning the co-dependency of living creatures on each other:

> Each oyster bed is a community of living creatures to a certain extent, a variety of species and the sum of individuals that can currently find the preconditions for their propagation and survival ... Science does not yet have a term for such a community of living creatures ... I call a community as such a biocenosis or living community.

Based on the oyster beds and human interactions with them, Möbius then developed the natural laws of ecological equilibrium, which had not yet been conceived as such:

> Although each species is organized differently, in each one different forces work together towards the creation and survival of the individuals; although each species has its own equivalent, nevertheless they all pose the same nutritional requirements to the entirety of the living conditions of the biocenosis.

As we know, human beings may have read the lesson that Möbius tried to teach us via the oyster, but we have hardly taken any consequences. If we are confronted with the effects of the anthropogenic climate crisis or the next large meteorite, the oyster will almost certainly survive, as it did all the previous catastrophes on our planet. Then the oyster would have committed its greatest feat—of surviving even the destructive forces of humankind. Human beings would have served the oyster, which they made a huge effort to protect and farm on account of the oyster's wonderful taste.

Humans are the oysters' keeper, their willing slave. That humans believe all their effort is to protect their own species—if we think along Richard Dawkins's lines—may instead be an evolutionary trick of the oyster's DNA, which knows exactly that it must sacrifice a few individuals to survive as a family. Every housecat knows what I am referring to here. The vision I had that afternoon at the Cambridge museum was that the oyster might even rule the world. Our whole planet might lie in the hands of a creature that managed to survive for a quarter of a billion years due to its good taste and tenacity; as well as its excellent, passive defense system; rather than an elevated level of intelligence or speed, fast reactions, or power of imagination. Its relationship with human beings is only a short episode, the blink of an eye, during its reign on Earth. Did Manet's *Beggar With Oysters* already know that? Or did the oyster even lend a hand in creating human beings?

Noble gourmands: Jean-François de Troy, The Oyster Dinner, 1735. Painting for a dining room at the Palace of Versailles, where empty oyster shells could apparently also be discarded on the floor.

KANT EATS AN OYSTER AND MEETS EVANGELINE

"How far such an one (notwithstanding all that
is boasted of innate principles) is in his knowledge
and intellectual faculties, above the condition of
a cockle, or an oyster, I leave to be considered."

JOHN LOCKE, *An Essay Concerning*
Human Understanding

HUMAN BEINGS DO NOT ultimately know much about our origins—despite all the efforts of science. We have gladly filled this void with stories ever since we had the power of speech— stories that sometimes give the oyster a leading role.

At the beginning of the ancient Greek schools of philosophy, Greek mythology revered the oyster as the origin of life. As I saw in the Cambridge University Museum of Zoology, that had a certain justification. It all began with the primeval soup, and the question is whether the chicken or the egg came first.

The primeval soup was the origin of shellfish, which gave birth to beauty. Mural in Pompeii in Casa della Venere in Conchiglia, depicting the birth of Venus (Aphrodite).

Aristotle answered this question clearly. It was not the chicken or the egg, but the oyster. In his *History of Animals,* Aristotle hypothesized that, unlike animals that reproduced sexually, oysters and other shellfish were spontaneously generated in the mud. Aphrodite herself was believed to have emerged similarly from a seashell.

A little later, in the Roman Empire, oysters almost lost their struggle with humans for world domination. No matter whether one believes the legend that Emperor Vitellius ate over a thousand oysters at one meal, Pliny the Elder reports that oysters were worth their weight in gold.

During the downfall of the Roman Empire and its culture, the demand for oysters fell and the natural populations in the seas recuperated. Soon the shellfish could be found everywhere once more. Not until the Renaissance did the demand increase again; at first, there seemed to be a wonderful coexistence of human beings and oysters. The creature was admitted to the most noble residences, and in 1735 it appeared in an official painting depicting a royal oyster banquet of noblemen at Versailles—the palace of the French King Louis XIV, known as the "Sun King."

It did not take long for the silent shellfish to become an object of philosophical observation. Its lack of mobility or sense of vision and hearing has made it an extreme form of life and a philosopher's stumbling block, as it is quite out of the ordinary. In his major tome *Émile, or On Education*, Jean-Jacques Rousseau postulated, "To the oyster the whole world must seem a point, and it would seem nothing more to it even if it had a human mind."

Even so, the oyster is not concerned with philosophical problems. It can't read Kant; however, Kant was able to eat it. And the most rational of all philosophers, who lived on the coast of the Baltic Sea, was a great oyster gourmet. The philosopher may not have appreciated their taste at first; however, they very soon became a focal point of his thought as something "irrational." In his work *Physical Geography*, Kant mentions them in a rather strange context:

> Oysters are often so firmly attached to a rock bank that they seem to be of a piece with it . . . They pinch with uncommon force when they shut and reproduce very quickly . . . One can also see oysters growing on trees as it were. These attach themselves to the branch of a tree in times of flood when the tree is under water and remain there.

Here—and I have always wanted to write this sentence—*Kant errs*, giving a misleading report of a land where oysters grow on trees.

Kant liked to philosophize about oysters. For this rational thinker of the Enlightenment, the consumption of shellfish was an example of the "hysteria" that takes hold of humankind when acquiring new tastes. An exaggerated refinement of the taste buds may dialectically result in a person elevating something to a delicacy, even if they may not have liked it at first. In his lecture on human science called "Menschenkunde," Kant postulated in the winter semester of 1781:

> In the case of our senses, there are many misconceptions, especially regarding that which we consider pleasant or unpleasant—for example, that every person must acquire a taste for eating the oyster, so that he will finally like the taste of the delicacy, this being based upon a recommendation. Thus, they believe it to taste good. Moreover, an oyster is healthy; although it does not initially taste good, a taste for them is only acquired after eating them for a long time.

However, Kant also became a victim of such conceptions himself. Although otherwise not necessarily inclined to sensual enjoyment, he did enjoy oysters. As a guest of the noble banquets held by the Countess Caroline von Keyserling in Königsberg, he often had an opportunity to observe such a transformation of tastes when eating oysters. "Four courses? That will suffice. But oysters, please," was his answer to one invitation. Kant does not explain his delight in oysters as something affected, rather, as a sign of an enlightened refinement of human manners. "The delight in the aftertaste is the greatest and best of all. Thus, one loves an old wine, and oysters, because their aftertaste is so good," he declared to his students during a 1772 lecture. As an ascetic, Kant only asked himself on occasion, quite theoretically, whether this indirect desire might also have a decadent tendency. From a practical point of view, he liked oysters, and so did Casanova. The latter preferred to slurp them from the tongue of his lover. One oyster-feasting scene, during which that philanderer seduced the Venetian nun M.M., is notorious:

> We amused ourselves eating oysters, exchanging them when we already had them in our mouths. She offered me hers on her tongue at the same time that I put mine between her lips . . . What a sauce that is which dresses an oyster I suck in from the mouth of the woman I love! It is her saliva. The power of love cannot but increase when I crush it, when I swallow it.

For the Italian historian of the sense of taste, Piero Camporesi, the oyster is a signature dish of the Age of Enlightenment, whether as libertine concept or idealistic philosophy. "Oysters and truffles seized power, forcing all the strong dishes typical of ancient aristocratic tables into exile." The oyster became bourgeois. "These glorious black and bloody meats suffered the affront of having to bow to the soft, bloodless and gelatinous pulp-like flesh of oysters."

It is not clear exactly when the parallel between feasting and orgies, between the oyster and the organ, first manifested itself.

*Deities, women, and oysters—all nude. An Olympic orgy:
The Feast of the Gods by Frans Floris.*

However, it was painted quite early in European art history. As soon as Renaissance artists began portraying the Olympic gods at orgiastic feasts, the oyster was used to symbolize the act that took place behind the screen.

Paintings of bacchanalian feasts without any oysters had already become inconceivable by the sixteenth century. The greatest orgiastic picture I know of was painted by the Flemish painter Frans Floris, who brought the Renaissance to Holland and from whom Peter Paul Rubens (and much later Andy Warhol) learned that a successful painter could employ an entire studio full of apprentices to satisfy the demand for art. When Floris imagines a bacchanal of the gods, there are cupids with red wings flying over the banquet table, and bearded gods with

Another bacchanalian feast, here with well-dressed guests in an earthly salon painted by Dirck Hals: Merry Company at Table *(1627-1629).*

nude chests leaning toward scantily clad goddesses. Mars, in the right-hand foreground, has laid his armor to the side and is amorously kissing his dining partner. Hercules, in the left fore-ground, has allowed his lion pelt to slip far down his hips. And the lady, who has won over the mightiest of the gods with her charms, is offering him an oyster as a form of prelude and invig-oration. Bacchus is serving wine, and oysters on the half shell are lying ready on the table in case anyone's libido should fail.

It is rare for the sexual connotation of the oyster to be depicted as openly as it was here at the beginning of the modern era of European art. The oyster's function as a prop represent-ing decadence remained unchanged when the persons depicted were not deities, but humans in a bourgeois interior. The orgy of the gods was transported back down to Earth a few decades later. The portrayals of the ladies and gentlemen on the paintings by Dirck Hals may not be as heroic or noble; however, their attire is all the richer and the atmosphere is equally as relaxed. Hals's painting illustrates the same theme. Life on Earth can be merry in intimate salons with plenty of ruffles on expertly tailored garments, and decadence may be impending... but the role played by the dish full of oysters amid the jovial diners has not changed in the least in comparison to that of the bacchanalian feast of the gods. Pleasure and lust have come down to Earth along with oysters on the half shell, all of which have made themselves at home in a bourgeois salon.

Slowly, however, the formerly aristocratic dish was to become a proletarian snack—cheap food for the masses. Strangely enough, the sexual connotation was never entirely lost amid the increasing "availability" of the oyster, which occurred due to the industrialization of its production. In the nineteenth century, many New Yorkers nourished themselves primarily with oysters. In a contemporary report, the Scottish journalist Charles Mackay noted with surprise regarding the not-particularly-genteel oyster bars that were the center of oyster consumption:

> The stranger cannot but remark the great number of "Oyster Saloons," "Oyster and Coffee Saloons," and "Oyster and Lager Beer Saloons," which solicit him at every turn to stop and taste. These saloons—many of them very handsomely fitted up—are, like the drinking saloons in Germany, situated in vaults or cellars, with steps from the street; but, unlike their German models, they occupy the underground stories of the most stately commercial palaces.

During this period, seafood had a career as cheap nutrition for the masses, and oyster production become industrialized in England, the United States, and France. In Baltimore, under inhuman conditions, women in factories filled thousands of oysters into tins each day, subject to the strict inspection of male inspectors. This mass production drove the oyster species to the edge of extinction once again. The proletarianization of the oyster almost became its demise.

By the nineteenth century, eating oysters had become so common that Charles Dickens quite astonishingly wrote, "It's a [v]ery remarkable circumstance…that poverty and oysters always seem to go together." The claim can be found in chapter 22 of *The Pickwick Papers*, published in 1836. In this book, the self-professed philosopher Sam Weller makes this statement and continues, "What I mean, sir…is, that the poorer a place is, the greater call there seems to be for oysters." Indeed, the journalist Henry Mayhew said in his publication *London Labour and the London Poor* of 1851 that around 500 million oysters (and around three times as many herring) were sold at one fish market in London in a year. It was during this period that Manet painted the portrait of the impoverished beggar-philosopher with several discarded oysters at his feet. Trains, able to transport the shellfish faster than they could die, were steaming to distribute oysters around the world at speed. Karl August Möbius wrote about this phenomenon in 1877:

> As soon as oysters could be shipped on steamers and trains fresh from the oyster beds, quickly transporting them deep into the interior of the land, ever more people began eating

A scene of gentlemen's bar culture by Richard Caton Woodville: Politics in an Oyster House, 1848.

oysters, annually increasing the demand, despite their growing prices. This took place almost as quickly as the growth of the railway networks in France, England, and Germany.

Oysters were an integral part of the street scene in the great, industrialized cities of the nineteenth century. The sale of the shellfish was often in the hands (and baskets) of so-called oyster girls, whose reputations were not always the best. Popular contemporary song lyrics reflected this:

As I was walking down a London Street,
A pretty little oyster girl, I chanced for to meet.
I lifted up her basket and boldly I did peek,
Just to see if she's got any oysters.

The girl promises the man free oysters. Then he rents a little room, where they spend less than "a quarter hour," before the oyster girl picks the pockets of the impertinent oyster lover.

These young women, who sold oysters as cheap snacks on the streets of metropolitan cities from the seventeenth century onward, ran the risk of being considered prostitutes. No differentiation was made between the salability of the oysters and the women, nor between the shellfish and a vulva: "The Oysters good! the Nymph so fair. Who would not wish to taste her Ware" read the caption of an eighteenth-century etching, *The Fair Oyster Girl*. In reality, however, the "oyster women" of the eighteenth century were not easy, young girls, but resolute women, as a rule, protected by several layers of petticoats against the smell of the shellfish.

An American oyster factory in the twentieth century.

The journalist Henry Mayhew interviewed a London oyster saleswoman in the mid-nineteenth century. She told him that poor people would not touch oysters, because the taste reminded them of snails. Those of her customers whom she preferred to call "gentlemen" would swallow down the oysters right at her stand as though they were taking poison, presumably because she was "down on her luck." However, her customers also included maids, who were sent to buy a fine dinner consisting of five or six dozen oysters.

The great minds of the era took the silent mollusk as an inspiration for metaphysical speculations. Charles Darwin, who repeatedly observed oyster beds on his journey around the world, and according to which he attempted to determine the

The **FAIR OYSTER GIRL** .

The Oysters good! the Nymph so fair, Who would not wish to taste her Ware

124 Printed for Carington Bowles, at Nº 69 in Sᵗ Pauls Church Yard, London.

Which wares are on offer here? The Fair Oyster Girl adapted from Philippe Mercier (printed between 1766 and 1799)—as an erotic fantasy.

In contrast: lower-class reality reflected in an old photograph of a street scene. To protect themselves from the smell of the shellfish and the water, the oyster women wore several layers of skirts.

age of the Earth, posed the rather absurd question of whether oysters have free will. In his unpublished notebook "Metaphysics on Morals and Speculations on Expression" dating from 1838, we can find Darwin's thoughts on the nature of free will. He attempts to understand the phenomenon with extreme examples, such as that of a baby—by nature irrational, thus supposedly not having a free will. One would have to assume the same of animals. Darwin explores this concept further by asking where the limit can be set. He avails himself of a most obscure creature, one which does not possess a brain. Could one presume that an oyster has a free will? "Now free will of oyster, one can fancy to be direct effect of organization, by the capacities its senses give it of pain and pleasure . . ."

Darwin postulates that because oysters react with their minimal senses to pain and pleasure, they can be said to have a free will. And it follows that "if so free will is to mind, what chance is to matter." For Darwin, chance was the actual motor of the evolution of life, the "natural selection" that he wrote about, which brought forth life in all its diversity. Yet just as the relationship between chance and matter organizes and arranges it into complex structures, free will likewise gives rise to the power of the mind. And thus, humankind did not evolve from apelike ancestors alone; instead, our minds originally stemmed from that of the oyster—at least our basic instincts can already be found in the primitive mollusk. This presents a great theoretical burden for a creature that does not even possess a brain! When we are confronted with Darwin's thoughts on the oyster, we are reminded once again of Manet's beggar-philosopher, who imparts his wisdom by the observation of oysters.

Back to reproduction, without which there would be no natural selection. It is always surprising that the oyster, perhaps one of the dumbest creatures, quite foreign to our species, is repeatedly used as an aphrodisiac. As a modern manifestation of the oyster girl, after World War II, the iconic exotic dancer Kitty West appeared as "Evangeline the Oyster Girl" at Casino Royale on the legendary Bourbon Street in New Orleans. She was born Abbie Jewel Slawson, purportedly a cousin of Elvis Presley, and she made her stage entrance in a giant oyster shell, dancing with an enormous pearl. Her burlesque act remains so well known to this day that the dancers of her dance school call themselves "oyster girls" and are proud to have been personally instructed by the inventor of this show number, who passed away in 2019 at the age of eighty-nine.

In her show, Kitty West exchanged the oyster's amorphic body for her own. One can imagine the reaction of the audience watching the show in the casino—perhaps holding a fresh-served oyster on the half shell in their hands. One oyster that became a woman before their eyes, and the salty body of another oyster in their mouth. To satisfy their urges, men, wearing their evening suits like hard shells, proffer bills of cash to the lightly dressed woman; the oyster also makes apparent the primitive drives of capitalism.

However, an oyster show could also be tender. At least that is how Michael Ondaatje describes it in his novel *Coming Through Slaughter*, which also takes place in New Orleans: an age-old

The human being as an oyster: Kitty West as "Evangeline the Oyster Girl" in New Orleans in the late 1940s.

oyster dance, where the oyster doubles as a symbol of physical intimacy:

> an Oyster Dance—where a naked woman on a small stage danced alone to piano music. The best was Olivia the Oyster Dancer who would place a raw oyster on her forehead and lean back and shimmy it down all over her body without ever dropping it. The oyster would crisscross and move finally down to her instep. Then she would kick it high into the air and would catch it on her forehead and begin again.

All right, time for a different approach to oysters, with some classical works of art.

STILL LIFE
WITH SHELLFISH

"All art is autobiographical;
the pearl is the oyster's autobiography."
FEDERICO FELLINI

EVER SINCE IT became the focus of my attention, I have noticed that oysters pop up out of the blue everywhere in the world. Although I originally worried about having to write excessively about classy restaurants or complicated methods of breeding, I became surrounded by oysters. No matter whether I was boating through the Venetian Lagoon, or hiking in the Karwendel Mountains, visiting Darwin's collection at the University of Cambridge, or sitting on a beach in the Caribbean, oysters were always in the vicinity. Naturally, I had expected to meet oyster lovers in the bars on the French Riviera, but not in philosophy. I knew that there were impressive portraits of oysters in Dutch still life painting, and that each of my next visits to museums would provide a renewed acquaintance with the little gray shellfish. Suddenly, the paintings were full of them. When

I went to the Alte Pinakothek in Munich, which is not necessarily famous for its nautical fascination, there were oysters everywhere, no matter whether it was Jesus preaching on a seashore, Jonah escaping from the whale, or a long-forgotten Dutch mayor serving his guests masses of them.

Rebecca Stott, who has written one of the best-researched books about oysters and their history, has observed that paintings of oysters began with large scenes observed as if through a wide lens, with the focus becoming closer and closer over the decades: "the 'camera' draws closer, lingering." The shellfish progressed from a casual requisite that accompanied a decadent feast to the center of focus in the drama. Yet the iconography of the oyster has hardly changed; it is always connected to erotic acts—young women offer mostly older men an oyster as a sign of their devotion, which may go as far as carnal lust.

In Frans van Mieris the Elder's tiny painting titled *The Oyster Meal*, which is hanging in the Dutch city of The Hague in a corner of the noble Mauritshuis museum, the oyster on the half shell doubtlessly symbolizes the female genitals. In this picture by a male painter, not necessarily famous for its beauty, a young, richly dressed woman with a deep décolletage is offering a rather unattractive man an oyster on the half shell. Under her open, fur-lined cloak, she is wearing an oyster-colored silk gown, giving an impression of a shellfish on account of the outward posture of her knees. The painting clearly has an underlying erotic vein. Between the scene depicted and the act that will soon take place on the curtained bed behind the two of them, the moist oysters on the half shell will be consumed.

The Oyster Meal by Frans van Mieris the Elder (1661).
Three creatures ready for anything—temptation,
seduction, consumption.

Perhaps a little sip of wine shall be had before the lovers draw the curtains?

At some point in time, the human subjects disappeared from these paintings, yet the oysters remained. If fresh seafood had before been nothing but a requisite on the table, it soon became the preferred prop in a novel style of painting. The painters only had to leave out the lovely lady and the lustful gentleman. Around 1600, the oyster became a model for the newly developed form of still life painting. That may seem surprising at first glance, since the oyster is extremely unsuitable for painting. No one would easily be able to paint an oyster using only the power of recollection—the way any child can draw a cow, a dog, or a fish. The oyster has an amorphous appearance. It does not have a tail or eyes, making it more easily recognizable. And it is not necessarily pretty, either. Perhaps the ill-defined shellfish provided artists a particularly good opportunity to demonstrate their virtuosity—rather than just painting the folds in gowns, shimmering materials, furs, carpets, curtains, glassware, and reflections, they could additionally portray the formless oyster in their pictures. At the beginning of the seventeenth century, the lowly creature, still alive on the plate full of inert objects, became the star of still life painting—known as *nature morte* in French, a more melodic name.

Seventeen years after Caravaggio had first painted a basket brimming with fruit in front of a beige wall, creating one of the first autonomous still life paintings, Osias Beert in Antwerp had already begun painting flowers, fruit, and glassware—as well as oysters—on tables. Not because they were particularly

Ten live oysters and a deadly fly are portrayed in the still life Bodegón by Osias Beert, circa 1600.

pretty, but simply because he was able to do it—and not intending to make a particular statement, but simply because these things were available. Their peaceful appearance was sufficient; no words were needed. These are the first paintings in the history of Western art that do not illustrate a story; instead, they are nothing but a picture. The shimmering of the mother-of-pearl; the moist, shiny appearance of the oyster's body; and the formless outer shell do not remind us of the battles of the gods, mythological matrons, or holy virgins. The uneven, shivering form of the raw shellfish creates a contrast to the exacting form of the handblown crystal wineglasses, traditionally placed next to the oysters in a still life painting.

These paintings can be interpreted by their symbolism, but that would completely contradict their essence, as they are simply an object of perception, nothing but time come to a standstill.

In his picture *Bodegón*, now in the Prado in Madrid, oyster virtuoso Osias Beert gives each of the ten oysters on the half shell its own individual shape. Beside the shucked shellfish, there is a fresh loaf of bread with a fly on it. At the time when still life paintings were interpreted in terms of symbolism, art historians contrasted Eros (the oysters on the half shell) with Thanatos (the mortality of bread, with a death-bringing fly) and inebriation (full glasses of wine) with the festivity of the moment (oysters that must be consumed shortly after shucking them). The fly, relevant in art history, may be laying its fatal eggs on the bread at the same moment, thus invoking the seed of mortality in this rich setting in the eye of the onlooker.

Yet this picture of the ten oysters, the wineglasses, and the bread with a fly on it does not revolve around the theme of mortality—quite the opposite! Beert has practically eliminated mortality by imbuing immortal objects with desire. The picture has captured time shortly before the moment of enjoyment. Both wineglasses are still full, and the oysters present a seductive sight for the onlooker. Their salty taste is perceivable in the picture. The first oyster has yet to be consumed. It makes me want to reach into the painting, shoo away the fly, eat the oysters, and drink the wine! However, the picture does not allow us to enjoy the delights that it promises. As René Magritte would say,

that is not an oyster; it is painted desire. A portrait of coveting—
the eternalness of lust was captured here. It is the creation of
desire, rather than the fulfillment of it. The enjoyment lies in the
eyes and the imagination of the onlooker. That is why there are
rarely any empty oyster shells in paintings, apart from in *Beggar
With Oysters* by Manet.

The American author Mark Doty has written an entire book
titled *Still Life With Oysters and Lemon: On Objects and Intimacy*,
where he makes a determined call to consider still life paintings
with oysters an extreme form of virtuosic art akin to love: "a
demonstration of virtuosity so extreme as to be explicable only
by means of love: this is a testament of falling in love with light,
its endless variation, its subtlety and complexity." Doty also
experienced an almost corporeal desire of the tongue while
looking at a picture by Dutch painter Jan Davidsz. de Heem in
the Metropolitan Museum of Art in New York, and as a poet, he
desired nothing more than to capture in his words the feeling
that this still life awakened in him: "Their liquidity makes me
want language to match, want on my tongue their deliques-
cence, their liquefaction." And then, in view of the lemon, the
wine, and the oysters, his words and memories flow onward for
the length of the book: "As if . . . the sharp pulp of the lemon, and
the acidic wine, and the salty marsh-scent of the oysters, were
some fragrance the light itself carried."

Intimacy and distance, presence and mortality, all visible in
a picture of a few immobile shellfish.

The titles of Dutch still life paintings usually reveal their
contents. One of the most beautiful is *Still Life With Oysters, a*

Rummer, a Lemon and a Silver Bowl by Willem Claesz. Heda. It was painted in Haarlem in the year of 1634 as one of the high-lights of still life painting, which was becoming increasingly monochrome. There is seemingly little mortality here. The setting of the silent drama is undetermined, as in the case in Caravaggio's masterpieces. A dramatic ray of light from the left illuminates the objects in the painting—including a rum-mer glass half-filled with white wine, which counts among the painter's standard repertoire; he must have had it stand-ing in his atelier, as he painted it in many of his pictures. The fallen silver goblet was also often integrated in his paintings. In the center of the picture, we see a lemon, which was a valu-able fruit in seventeenth-century Holland. It is present in each of Claesz. Heda's paintings and is always represented in a slightly different form. In the present picture, its rind has been partially peeled. It hangs loosely in bright yellow off the edge of the table, as though it might escape from the picture. This method of painting a lemon lying beside the oysters was to endure over the coming centuries. The unraveled lemon rind, which is difficult to paint in its complex geometry, became a subject of art history, as did the oyster itself. "Lemon com-petition" is the name Mark Doty gave to this commonplace appearance of virtuously depicted lemons in art history. Two quite plump oysters are lying beside the lemon. Here they will be spiced with salt and costly pepper, which can be seen in a small paper twist located in a precarious position on the edge of the table, as though it might fall off the picture—just like the lemon rind.

In the greatest idyll, decadence may lie, as seen in this Still Life With Oysters, a Rummer, a Lemon and a Silver Bowl *(1634)*.

It would seem to depict a good life, but only at first sight. A closer look at the strictly triangular ensemble reveals another glass on the left-hand side of the table, the broken shards of which have fallen into the dish of oysters. This symbolizes mortal danger. The observer is torn between the attractive oysters and the repellent shards of glass. And once the first idyllic impression has been shattered, a person takes a closer look. The reflection of a window crossbar is visible in the half-filled rummer. It is located exactly in the center of the picture. Suddenly it appears uncannily like a threatening crucifix. Yet how can that be? In the larger, clearer reflection of the window on the upper rim of the glass, one can see that the window, through which light filters into the room, does not have a crossbar! Only the two oysters, the sole living elements in this *nature morte* painting, are more radiant than the sources of light or the silver bowl. These are the painter's actual masterpiece, tempting with their soft flesh on shimmering mother-of-pearl. The oyster liquor is so clearly visible in the shells, as though it were to begin dripping off the dry canvas of the painting. If this still life were a musical composition from the late 1960s, it would be a never-ending guitar solo; we are captivated by the vision.

The depiction of oysters in such a realistic manner makes one's mouth water. Because of this, Claesz. Heda's still life paintings are one of the main attractions in the Rijksmuseum in Amsterdam, for which tickets must be reserved long in advance. The museum is far from an oyster bar, where tasty Dutch Zeeland oysters are available, grown nearby in the cold North Sea waters of the Oosterschelde. Go to the museum. In this huge exhibition hall, you will not be able to avert your gaze from the two oysters in this painting, which have been waiting to be eaten for almost four hundred years. Strangely, the oysters are the only mortal beings in the picture. They would not have survived beyond the time of their depiction. The painter's servant likely had to keep them alive with salt water while they were being portrayed, just as is done in contemporary food photography, because they would not have retained their sheen for the entire time it took to paint them. Yet you will immediately want to eat these oysters and taste the chilled wine in the picture. Art cannot be more hedonistic. You could quickly wipe your hands

The beginning of the modern age: Luncheon in the Studio *by Édouard Manet (1868). Only the oysters on the table look the same as always.*

on the tablecloth before the museum guards appear. However, you would not touch the valuable silver goblet. The museum can keep it.

The Dutch oysters with lemon became a leitmotif in art history from this point onward. The iconographic painting *Luncheon in the Studio,* completed almost 250 years later in 1868 by the young rebel Édouard Manet—considered a vanguard of modernism—alludes to Dutch still life painting. A few years earlier, Manet had painted the beggar-philosopher with oysters in a more academic style. At that time, the world of art was somewhat out of kilter. Impressionism had its focus on the moment, rather than on history. On the right hand, below the table where the young man is leaning, you can see Dutch still life painting reflected in the curling lemon rind, a huge oyster, and the obligatory knife pointing over the edge of the table. The man at the table, who is his fellow painter Monet, has presumably just eaten a few oysters. Time is standing still here too. The three people also seem to be frozen like the objects in a still life painting, shifted slightly toward the edge of focus.

A few years later, when Manet had already become a successful Impressionist, he painted a still life with oysters once again, as this style of painting was considered classic. In *Nature morte, huîtres, citron, brioche*, he zooms in for a close-up portrait. Down to the tablecloth, his arrangement reflects that of his Dutch colleagues—a few oysters, lemon, and bread without a fly. The picture is a scholarly citation; the artist has painted the timeless theme of oysters in his own contemporary style, reflecting art history and then laying claim to his due place in it. Manet seems to sense that a single stroke of the brush will soon be considered a work of art. Therefore, he serves us oysters to prove that his own modern art can be compared with the greatest of his colleagues and those of the previous centuries. In his painting of a dish of oysters dating from 1876, the shellfish prepared for consumption are lying in front of a naked wall—yet the table is covered with a cloth and there is a knife on it, along with bread and butter. As in the painting by Willem Claesz. Heda, the light is falling from the upper left-hand corner, yet the illumination is not given much attention by the

An elegantly simple meal. Nature morte, huîtres, citron, brioche by Édouard Manet—the usual oysters, lemon, bread, and butter—and a few brushstrokes.

painter. There is no symbolism involved here. The lemon has simply been sliced through its middle, and the rind is lying in the background. A fast meal, a quick painting, a profane butter dish instead of a noble goblet. Manet proudly shows that he can portray eight oysters much better, with far fewer brushstrokes, than his predecessors could. Welcome to the Modern Age! Observers' eyes create the world with the brushstrokes that primarily reflect themselves. They are full of color. The celebration of the moment has passed on from the general theme to the picture. Manet's painting is intended to give the impression that it was completed quickly after the oysters were shucked,

Georges Braque heralded the end of art history in commerce: *Oyster and Lemon (1953), invitation card from Galerie Maeght.*

yet before they were consumed. The painter likely ate his models after his painting was finished. This could have marked the end of art history and the end of oysters in it; however, it was only the second-last chapter.

The avant-gardists of the twentieth century were still interested in how a strange creature like an oyster could find its way into a picture. *Oyster and Lemon* is the title of Georges Braque's painting done decades later, his version of the classic theme on the invitation to a gallery exhibition. The tablecloth, only the pattern of which is similar to Manet's, is already as abstract as the future will be. No one wants to eat this oyster anymore. It is not particularly appetizing, nor desirable. It is just a picture, sufficient in itself, quite elevated beyond the boring, realistic world. This oyster is nice to look at. In fact, it simply represents a distant recollection of a formerly desirable object. It does not invite us to enjoy a meal. Instead, it invites us to buy works of art, or even to speculate with them. Braque's lithography decorated the invitation sent by the influential French Galerie Maeght to a few rich collectors, who would collectively create art history from there on.

A HAPPY END

"No, I do not weep at the world—
I am too busy sharpening my oyster knife."
ZORA NEALE HURSTON

THE EUROPEAN FLAT OYSTER, which already populated European coasts before its contemporary cousins—also serving well to nourish Stone Age people—has already been practically extinct for almost a century. A proven fact. Industrial overfishing, exploitation with dragnets, and the destruction of the environment have almost completely wiped it out. Not a single oyster can be found in their former habitat of the North Sea or the Atlantic coast of Europe. A few tiny remaining oyster beds, carefully tended by oyster farmers in Norway and England, can still be found. However, there are none left to speak of on the Moroccan coast, or in the Mediterranean or the Black Sea. "The European oyster is considered functionally extinct on the German coast of the North Sea," stated the German Federal Agency for Nature Conservation in its literature on the restoration of the populations of the European flat oyster (*Ostrea edulis*). Yet the pioneering environmentalist Karl August Möbius already made a fundamental declaration in 1877: "The preservation of oyster beds is as much a question of statesmanship as the preservation of forests." Time for a happy development. Enter the marine biologist. For my generation, growing up with TV broadcasts of Jacques Cousteau's adventures, "marine researcher" was the job of our dreams. Today, his contemporary scientific counterparts in Germany work at the Alfred Wegener Institute. Since 2017, a team of researchers under the direction of marine biologist Bernadette Pogoda has been working on the reestablishment of the European oyster in the North Sea. In the photos of her institute, you can see her seated between test tubes or wearing a life vest onboard an expedition ship. Pogoda has become a mother to hundreds of thousands of oysters. However, her job is a far cry from romantic research at sea. On the phone, she speaks so quickly, it is clear she knows that time is of the essence.

You might think it would not be overly complicated to re-establish indigenous oysters. It should only involve putting a few of the creatures, which have lived here for millions of years, back into the sea. Yet the world has become a complicated place. In the twenty-first century, it was not even simple to fulfill the legal stipulations for this important project. In 2011, the so-called shellfish judgment passed by a German administrative court strictly prohibited the distribution of shellfish spat in national marine parks. It makes no difference whether the shellfish are indigenous or not. Before the praiseworthy project could begin, complicated legal assessments had to be made. In view of this, the image of the brave marine biologist weathering the windy elements on a cool research boat quickly fades. Pogoda's struggle turned out to be a complex process of authorization.

Only after the lengthy process of approval had been completed could her team begin their work. A few miles away from the island of Borkum, in the vicinity of a wind turbine park, a total of twenty-four thousand seed oysters, measuring two millimeters each, from eight baskets, were attached to the seafloor. Breeding these seed oysters was not simple. You cannot buy extinct shellfish at a market. The few indigenous oysters available from commercial breeders were not suitable for reestablishment in the wild, because these are genetically as similar to each other as possible, to provide similar oysters for the table. To reestablish indigenous shellfish, the greatest possible biodiversity was required—the largest genetic diversity to ensure that a variety of oysters could adapt to the numerous demands of life in the North Sea. In addition, the creatures to be reestablished in the marine park had to be "clinically purified" to avoid introducing new invasive species, worms, or parasites to the North Sea—as was the case in the 1970s in the Netherlands. Therefore, the Alfred Wegener Institute built its own laboratory on Heligoland for the purpose of breeding oyster larvae with ultramodern propagation methods, certified annually, to breed the practically extinct European flat oyster. The oyster parents came from France, in the estuary of the Belon River, as well as from Scotland and Norway—those regions where tiny oyster beds of indigenous oysters still existed.

First, "mature" oysters were introduced to artificial salt water in quarantine on land in a specially built breeding

Shortly before their near-final extinction:
European oysters (1870).

laboratory, held at suitable temperatures in plastic containers, and provided with a microalgae diet, encouraging them to spawn. Even the breeding of the sterile microalgae feed for the oysters was a science in its own. The larvae were bred in large test tubes, then moved to plastic tanks in an oyster house, where they were attached to old oyster shells, a method called "spat on shell." Then they are fed in their breeding tanks until they reach a size of two to three millimeters, and finally taken out to form a new oyster bed in the North Sea. Larger oysters often cannot be acclimatized in the wild due to the risk of disease or invasive species. As per government regulations, the first step involved a pilot-project oyster bed in Borkum, so a few baskets of carefully bred juvenile shellfish were introduced to the sea. As hoped, after nine to fourteen months the mollusks had reached sexual maturity. They had become males, later transforming themselves into females. As predicted, small oyster beds formed. NORA, the Native Oyster Restoration Alliance, was founded. Its goal is to support the creation of new beds of European oysters, growing independently. "It has been a great success to date," Bernadette Pogoda gladly confirms. Her job is secure in the future. The nature protection zone encompasses 241 square miles, where researchers from the Alfred Wegener Institute are currently at work. Around 100,000 oysters have already been acclimatized in the wild, a tiny number considering the size of the nature protection zone (and the size of the North Sea). Nevertheless, the continuation of the project has already been mandated by law. Will we be able to count on finding wild European oysters for sale again in fish markets in the future? Bernadette Pogoda thinks that might come to be. Naturally, however, all fishing activities are forbidden in the nature protection zone of the Borkum Reef. For now, she hopes that the oyster population will stabilize and form its own habitat. Ideally, the shellfish will multiply so well at a depth of seven to nine feet that the oysters will spread beyond the nature protection zone, so that someday in the distant future, classic oyster fishing of wild shellfish from the seabed will become possible once more.

This is urgently required in other locations, as well. The complicated project at the Borkum Reef has been replicated by nature itself, a few miles to the northeast. The tidal shallows around Sylt prove how quickly oysters can multiply, as wild oysters have meanwhile become plentiful in the region.

There, however, it is not the European oyster that has multiplied, but the Pacific oyster, which has populated the area so quickly that it has practically been deemed a plague. When the indigenous European oysters died out around 1900, there were various attempts to acclimatize other species there. All those trials were aborted in the twentieth century. Instead of oysters, nothing but slipper shells and American oyster drills spread here, both of which are considered invasive species. During the 1970s in the Netherlands, attempts were made to acclimatize diverse species of oysters in the wild. Parallel to this, German fishery researchers tried to get oysters to survive on the German coastline of the North Sea. None of these attempts succeeded. It is interesting how often human beings can err and how nature reacts to these failed attempts. Soon the imported Pacific oysters were multiplying in the Oosterschelde, despite

THE PISCATORIAL ATLAS.

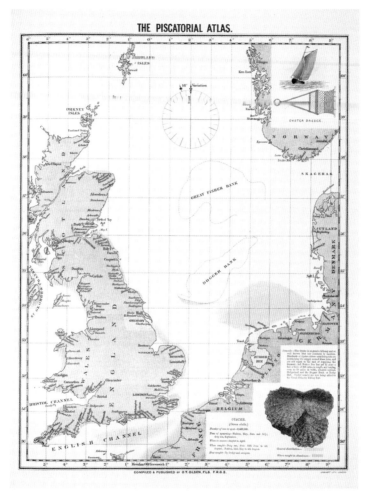

Former habitat of the indigenous European oyster in the North Sea and Atlantic Coast.

all contrary convictions. From there, they spread as an invasive species throughout the entire North Sea. No one can stop them, and almost no one can harvest them. Around the island of Sylt, these "wild" oysters are now in the process of displacing the indigenous blue mussels, in competition for their nutrition and habitat. Large Pacific oyster beds have already established themselves there. Climate change is warming these waters and enabling the spread of the Pacific oyster northward.

Considering their immobile state, the talk of "wild" oysters may sound strange. The idea has a romantic tinge, as if the freedom-loving oyster were able to make its way into the wild with the help of its sharp shell, as though it could drift in the ebb and tide in a "wild" state, living a long life outside of the fine-meshed nets of the oyster farm. However, so-called wild oysters have become an uncontrollable scourge. They are found in great abundance around Sylt, as well as in Brittany and in the Venetian Lagoon. There they are growing so quickly that they have already disturbed the circulation of the water in the Lagoon. Like their counterparts, growing under supervision in oyster farms for people to eat at dinner, these oysters do not really belong in those waters. They are invasive Pacific oysters— the progeny of those pioneers imported as seed oysters in sacks and boxes in the 1970s from Japan to France, and then brought to Germany in 1986. They actually should have been part of the human food chain after three years on an oyster farm. It was never planned that they live in the wild, nor was it considered likely. But mysteriously, these "wild" oysters appear to be more "freedom-loving" and resistant to disease than their farmed relatives in plastic nets on oyster beds. They are not even affected by the frequent epidemics that regularly drive the oyster farmers to the edge of ruin.

These oysters have replaced the blue mussels on a wide scale in this area. This might seem like good news for oyster lovers; yet it is fatal for some animal species, such as eider ducks, which cannot open oysters, in contrast to the indigenous blue mussels in their habitat. It seems the Pacific oyster is the first invasive species to negatively affect the tidal shallows here. In the meantime, fishers and oyster farmers have been given permits to collect and sell the invasive Pacific oysters from the tidal shallows. Sylt-based Dittmeyer's Oyster Company, the only German oyster farming enterprise, takes a boat out almost weekly to the large oyster beds. Only individual oysters are collected; however, most of them have grown together into large messy clumps, difficult to handle and impossible to separate, even with a hammer. The invasive species has achieved what the indigenous species could not do—it has formed its own oyster beds, where the species can multiply without constraint and remain unavailable for commercial use. This is its first conquest over the human being.

Those few oysters collected by oyster fishers are only briefly purified. To sell them commercially, a lot of forms must be filled out, detailing where, when, and how many oysters of which species have been collected. Eventually the "wild" oysters can be served in a restaurant as an ideal "connection between the former practice of oyster fishing and our traditional oyster farms."

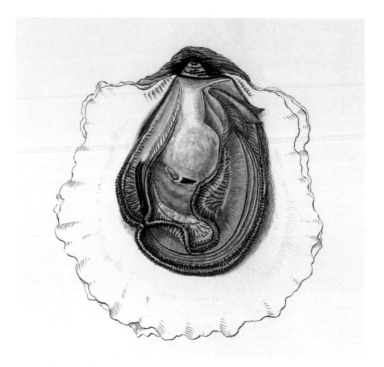

Human beings are making a huge effort to reintroduce indigenous species they drove to extinction, thereby testing the limits of legal regulations. A European oyster, slightly idealized for the portrait.

At least, that is the advertising copy. Of course, there is nothing traditional about this, and oyster fishers "of old" did not collect an invasive species from the sea. Instead, they dredged European oysters from the seabed until they became extinct. The population of Sylt mostly prefers the indigenous oysters from the oyster breeders, instead of the stronger tasting "wild" oysters. The few oysters collected from "wild" oyster beds do not alleviate the problem caused by the huge invasive population that already lives there. Private persons are officially forbidden to collect oysters. Only the holder of a fishing license may collect one three-gallon pail of oysters a day.

Nevertheless, there are people who go out at low tide to collect oysters for their consumption. But even among brave oyster collectors, there are reservations about gathering the huge wild creatures called *pieds de cheval* (horse hooves) and eating them raw without any purification. The practice of consuming wild oysters is not very widespread. It requires a rather wild spirit. "I have never seen anyone else here on these cliffs with an oyster knife and lemon," said the German author Christoph Peters about his beachcombing in Brittany. "There are oysters in all sizes ... I usually break off the extra-large ones, because they have twice or three times as much flesh in them, in comparison to a normal 'party oyster.'" And then the writer details the tale of his first "wild" oyster in his letter, his Oyster Conversion Experience:

> I coincidentally "discovered" them almost exactly twenty years ago. I thought those "flat" shells on the cliffs look like oysters, so I took a look ... Then I shucked one with a Swiss knife and gave it a try; it tasted like an oyster too, and I filled my belly. That evening, Veronika told our landlady, an old Breton widow, who was worried I'd poisoned myself ... and Madame Tourbon smiled from ear to ear. Her deceased husband often went out at ebb tide with an oyster knife and a lemon and ate them right off the cliff.

I must admit that I would not have been that courageous, myself. However, I did find my own wild oyster, not in the clear waters of the oyster beds around Brest, but in the brackish water of the Venetian Lagoon. During a boat trip in the Grado Lagoon, south of Venice, I got stranded on a sandbank at low tide and accidentally rammed the boat hook into the middle of an oyster. The magnificent shellfish had been living undisturbed on the sandbank for five or six years, judging by its size. While trying to measure the depth of the water, I had stuck my boat hook into an open female oyster that reflexively shut, gripping my boat hook tightly. I was surprised to see it at the end of my boat hook as I pulled it out of the water. That was the first time I came face-to-face with an oyster, as it were, outside of a fish shop. For hours, I proudly observed the only oyster I had ever gathered in my life, until the tide came back in and set our boat afloat again. Before that I had not realized that oysters had repopulated the Venetian Lagoon. Due to the legendary pollution in the Lagoon, making swimming hazardous, yet in which oysters apparently thrive, I did not dare to eat the freshest seafood I had ever found. I removed the flesh of the wild creature and threw it with a sigh into the Lagoon, which is still filled with wild, yet inedible, oysters. The boiled shell is lying beside me during the writing of these pages.

I asked Bernadette Pogoda about the significance of "wild" oysters in terms of her repopulation project. Is there any competition in the tidal shallows between the aggressively spreading Pacific oysters and the carefully protected European oysters? "Their habitats are quite different," she explained to me. "The European oyster mainly dwells in a sublittoral habitat," meaning far below the surface of the water, and it is not an invasive species that creates havoc on the array of indigenous species. On the other hand, the Pacific oyster grows in the tidal zone, where it is washed over daily by the flood tide. The two species can live side by side. The Alfred Wegener Institute is taking care that the European oyster can only repopulate areas where it historically lived. Their population has grown to around 100,000 specimens, and they serve to filter the water in the protected reefs where they live. Hopefully, the European oyster will attract other indigenous marine species to share its newly created, formerly existent habitat.

And then nature will be restored to the way it was in the times of Aphrodite, before people appeared with all their dragnets, oyster bars, and anchors. Oysters will be found everywhere, not only in restaurants and plastic nets. They will nourish, embellish, and stimulate humankind, as always. Their future is bright. The Earth will again become a planet of oysters and seas, and even the Lagoon of Venice will be cleansed by them, once all the people are gone.

PORTRAITS

PACIFIC
(GIANT) OYSTER

Magallana gigas

ALTHOUGH IT HAS only become established in Europe over the past half century, hardly any other species of oyster has any notable population on the European coastline. This shellfish has many names according to the country it populates, such as *huître creuse* in France, Sylter Royal in Germany, and Irish rock oyster in Ireland. They are simply known as Pacific oysters in the United States and Canada—when they are not called various fantastical names by the different oyster breeders.

At the beginning of the twenty-first century, they constituted over 95 percent of the world oyster production. Since the almost complete extinction of their related species, such as

Ostrea edulis, this incredible Asian oyster has become the preferred species bred by oyster farmers.

Magallana gigas looks like we expect an oyster should; however, that is also the result of breeding. In the wild, these oysters take on almost any warped shape imaginable, depending on the space available. The lower, "left" half shell of the entirely asymmetrical mollusk is always highly curved, while the top, "right" top half shell is flatter—yet still slightly convex. The shells, which usually assume a teardrop shape, are hard with sharp edges. The creatures themselves mostly have visible dark cilia. In China, particularly large oysters are bred. Almost four million tons of oysters grow there annually, which is equivalent to over 80 percent of the world production.

EUROPEAN
FLAT OYSTER

Ostrea edulis

THE LUXURY EDITION among the luxury dishes. From a culinary point of view, the European flat oyster is the greatest competition for the highly prevalent Pacific oyster. Because it is much rarer, it is considered an expensive delicacy, tastier than the "flabbier" Pacific oyster. In contrast to the more common oysters of the *Magallana* genus, the *Ostrea edulis* usually conforms to a more orderly, almost symmetrical round shape. Its appearance is closest to the legendary shells in depictions of Aphrodite's birth. The *Ostrea* is flatter than its more common competitor, which gave rise to its English and French vulgar names. Although it is called flat, its left and right half shells are convex.

Its shell is a lighter color than its fatter cousins—sometimes almost beige—the flesh is firmer, more muscular, and often an almost light brown color, whereas its taste is a little "nuttier" than its relatives, which grow at greater depths.

Although the *Ostrea edulis* was originally at home in Europe, it is now only grown in a few oyster farms. A focal point of industrial oyster breeding is the estuary of the French Belon River, which is why these mollusks are often sold as Belon oysters. In Ireland, they are called according to their farming region in Galway. In England, these oysters are often still traditionally fished from the seabed and simply called "native" oysters. In contrast to the other species of oysters, the eggs are fertilized in the shell of the mother oyster, which holds them back between her gills.

EASTERN OYSTER

Crassostrea virginica

IF WE CAN TRUST the world market, then the Eastern oyster is the second most important species of its kind. It only has a global market share of around 5 percent. As can be expected from its name, it grows on the Canadian and American East Coast down to the Gulf of Mexico, where it is still found in the wild, and in the Caribbean, where it is sold as a wild shellfish by fishers on the beach. Eastern oysters are oval, and the left half shell is convex, but not as highly curved as its European and Asian cousins. Its entire appearance is slightly less wild, almost elegant. These shellfish may have played an important role in American colonial history. One of the first English pioneers,

John Smith, likely admired these oysters in Chesapeake Bay in 1608. The famous Powhatan interpreter known as Poca-hontas possibly enjoyed eating them herself. In the winter of 1609, when many Jamestown residents starved to death, some of the first colonists survived only by nourishing them-selves for months from the rich oyster beds that formerly populated the American East Coast. In the nineteenth century, New York was thought to have been built on Eastern oysters. In the words of the American journalist Mark Kurlansky, "When people thought of New York, they thought of oysters." Those were the days.

KUMAMOTO
OYSTER

Magallana sikamea

THE KUMAMOTO, or "Kumo" for short, is the noble little sister of the Pacific oyster. It also originated in Japan, to be more precise in the Ariake Sea, a bay on the coast of the Kumamoto Prefecture. It is at most half the size and rounder than the former, and its taste is more subtle and refined that the stronger Pacific oyster. Its lighter flesh often tastes pleasantly "buttery," despite its high mineral content. It has become particularly popular in the United States, where it is like the indigenous Olympia oysters, which have almost become extinct.

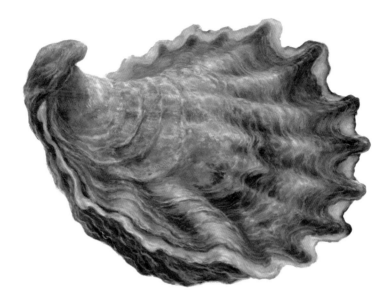

The Kumamoto is the perfect oyster. However, farming it is very time-consuming, as this species achieves a comparably low weight, so it is not the oyster farmers' favorite shellfish. Its shell often has a variable brown to greenish coloring and usually exhibits a visible, attractive ribbed pattern. The oysters themselves are not very demanding and are farmed primarily on the American West Coast, where they are considered a Californian delicacy on account of their almost buttery taste; they are less well known in Japan, their country of origin. The Kumamoto does not play a role in Europe either.

PORTUGUESE OYSTER

Magallana angulata

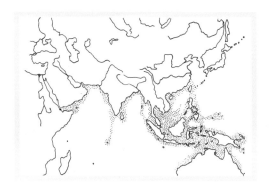

UNFORTUNATELY, THE CLASSIFICATION of this oyster does not follow the uniform terminology of the rest of the oyster family. In addition to the Latin name *Magallana*, there is some skepticism about whether the Portuguese oyster actually belongs to a separate family.

Its phenotype is like the Pacific oyster, from which it can hardly be differentiated. Irregularly formed, with a deep bottom half shell, it has an almost flat top shell. Genetic analyses have determined that it is a separate species; however, it was not indigenous on the west coast of the Iberian peninsula, as indicated by its name. Instead, it was imported from Asia by colonial

trade ships in the sixteenth century. Before it almost became extinct on account of an epidemic in the late 1960s, it had been the favorite and most popular species of oyster in Europe for over a century. Following a shipwreck in 1868, the so-called Portuguese oyster displaced the formerly leading *Ostrea edulis*, first on the Atlantic coast of France and then on the plates of oyster lovers all over Europe. However, history's revenge led to the faster-reproducing and even-faster-growing Asiatic *Magallana gigas* nearly replacing this invasive species within one hundred years.

EUROPEAN
THORNY OYSTER

Spondylus gaederopus

AMONG ALL THE OYSTER SHELLS, which can often be a little ugly, European thorny oysters can be considered real beauties. Their spiny shells and their usually colorful appearance were so attractive to Stone Age people that the empty shells were traded over great distances. It has been documented that these spectacular shells were used as jewelry by Neanderthal people around fifty thousand years ago.

These oysters originated in the Black Sea and the Mediterranean. (There is also a species called the Atlantic thorny oyster in North and Central America—*Spondylus americanus*.) Their lovely shells continue to amaze even serious researchers, who

speak of the most beautiful mollusks in the world. In contrast to other oyster species, both half shells are concave. This is possible because the shellfish only attaches itself to the seabed in one spot, near the hinge of its shell.

A spectacular appearance can also be a jinx for shellfish. In terms of evolutionary history, the decorative spines are not intended to beautify the oyster or defend the pretty creature from predators or shell collectors. Instead, their shell provides a home for other marine organisms to attach themselves, thus enhancing the camouflage of the attractive creature. The deeper the habitat of these oysters—they grow at a depth of up to 165 feet—the more colorful their shells, from flaming red to shimmering white to violet, or almost any other color.

PACIFIC
PEARL OYSTER

Pinctada margaritifera

IN CERTAIN CALIFORNIAN souvenir shops, you can pick a
live pearl oyster from a bowl for a few dollars. The salesperson, assuring you that each of the oysters really does contain a
pearl, will shuck it and present you with the pearl—after removing the otherwise valuable flesh, which is inedible in this case
(pearl oysters are not closely related to the edible oysters in the
Ostreidae family). The result will rarely be as nice as the most
storied, famous pearls.

Legend has it that Mark Antony, or Marcus Antonius, the
Roman general, vied with the beautiful Egyptian queen Cleopatra to hold the most extravagant feast—yet he found nothing

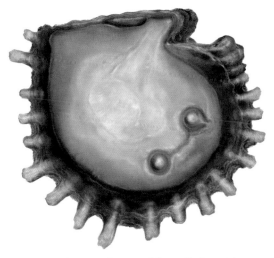

but two empty plates and two goblets of wine at her erotic bacchanal. In a scene that has become iconic, Cleopatra took off her pearl earring, dissolved it in her wine, and then drank it, thus winning the competition. The value of this pearl, known as "teardrop of the gods," was estimated by the Roman historian Pliny at 10 million sesterces, whereas a legionary's daily wage was 2.5 sesterces at the time. Contemporary scholars have attempted to verify this tale; however, pearls cannot be dissolved in wine, not even in vinegar.

Elizabeth Taylor starred as Cleopatra in the 1963 motion picture partially based on this tale. Richard Burton lay at her feet, and while they shot the movie, he fell in love with the diva, who had already been married four times. Later, he gave Liz Taylor the most famous pearl in the world for Valentine's Day, the pear-shaped pearl La Peregrina, whose name in English means "the Wanderer," and which according to legend had been jinxed.

ACKNOWLEDGMENTS

I WOULD LIKE TO THANK Elke Link, who, since once claiming she shared half an oyster at the Atlantic coast, has also shared her life with me. Moritz and Felix, not only for being oyster lovers, but also for being able to shuck them. The "Oysterband" for stoically bearing my attempts to prepare oysters. Kat Schumacher, who told me about foxes, which gave me the idea that oysters are much foxier; and Dr. Rudolf Pospischil, who kindly claimed that this book is almost as interesting as the contracts that he otherwise has to read.

FURTHER READING

Bottura, Massimo. *Never Trust a Skinny Italian Chef*. Phaidon, 2014.

Bourdain, Anthony. *Kitchen Confidential: Adventures in the Culinary Underbelly*. Bloomsbury, 2000.

Camporesi, Piero. *The Anatomy of the Senses: Natural Symbols in Medieval and Early Modern Italy*. Polity Press, 1995.

Casanova, Giacomo. *History of My Life*. Everyman's Library, 2007.

Darwin, Charles. *Notebook M* [Metaphysics on morals and speculations on expression]. Edited by Paul Barrett and John van Wyhe. *Darwin Online*, (1838) 2009. https://darwin-online.org.uk/content/frameset?viewtype=side&itemID=CUL-DAR125.-&pageseq=1.

Doty, Mark. *Still Life With Oysters and Lemon: On Objects and Intimacy*. Beacon Press, 2001.

Dumas, Alexandre. *Alexandre Dumas' Dictionary of Cuisine*. Edited, abridged, and translated by Louise Colman. Routledge, 2005.

Fishe Shelly, A. [James Watson Gerard]. *Ostrea; Or, the Loves of the Oysters.* T.J. Crowen, 1857.

Fisher, M.F.K. *Consider the Oyster.* North Point Press, (1941) 1988.

Harris, Stephen. "Oyster Rules Are There to Be Broken." *The Telegraph,* July 18, 2015.

Hemingway, Ernest. *A Moveable Feast.* Scribner, 1964.

Jacobsen, Rowan. *The Essential Oyster: A Salty Appreciation of Taste and Temptation.* Bloomsbury, 2016.

Jacobsen, Rowan. *A Geography of Oysters: The Connoisseur's Guide to Oyster Eating in North America.* Bloomsbury, 2007.

Kant, Immanuel. *Lectures on Anthropology,* Cambridge Edition of the Works of Immanuel Kant. Cambridge University Press, 2012.

Kant, Immanuel. *Natural Science,* Cambridge Edition of the Works of Immanuel Kant. Cambridge University Press, 2012.

Kurlansky, Mark. *The Big Oyster: History on the Half Shell.* Ballantine Books, 2006.

Mayhew, Henry. *London Labour and the London Poor.* Penguin Classics, (1851) 1986.

Ondaatje, Michael. *Coming Through Slaughter.* Vintage, (1976) 1996.

Redzepi, René. *A Work in Progress.* Phaidon, 2013.

Rousseau, Jean-Jacques. *Émile, or On Education.* Penguin Classics, (1762) 1991.

Stott, Rebecca. *Oyster.* Reaktion Books, 2004.

Wilson, Andrew. *Beautiful Shadow: A Life of Patricia Highsmith.* Bloomsbury, 2014.

ILLUSTRATIONS

Page 5. George Shaw, *The Denticulated Oyster*, in *The Naturalist's Miscellany*, plate 675, 1813.

Page 10. Édouard Manet, *Beggar With Oysters (Philosopher)*, circa 1865. The Art Institute of Chicago, Illinois.

Page 17. François Jules Pictet, *Ostrea rarilamella* (detail), in *Traité de paléontologie*, 1853–1857.

Page 20. Leopold von Schrenck, *Ostrea in Natural Form, Sketch on Stone*, in *Reisen und Forschungen im Amur-Lande in den Jahren 1854–1856*, 1858.

Page 25. Alfred Edmund Brehm, *Oyster, Opened by Removing the Top Shell*, in *Brehm's Life of Animals*, vol. 10, 1893.

Page 28. James Pradier, *Venus With a Shell*, daguerreotype of sculpture taken by Salvatore Marchi, 1851–1855. Courtesy of Stadtgeschichtliches Museum, Leipzig.

Page 32. Anonymous, *S.T.A.I., La Perle*, postcard, circa 1930.

Page 36. Sherman Foote Denton, *Oysters, Natural Size*, chromo-lithograph, 1902.

Page 38. Liebig's trading cards, *Les huîtres 1–6*, 1954.

Page 41. George Brettingham Sowerby, *Plate I: Ostrea*, in *Conchologia iconica*, 1843–1878 (1871).

Page 43. H.M. Smith, *Starfish Attacking an Oyster*, in *Annual Report of the Commissioner of Fisheries*, Washington, DC, 1921.

Page 45. W.G. Mason, *The Oyster Stall*, in Henry Mayhew's *London Labour and the London Poor*, 1851.

Page 48. Anonymous, *Vessels Dredging for Oysters*, in *Popular Science Monthly*, vol. 6, 1874–1875.

Page 50. Sylvanus Hanley, *Ostrea sinensis of Gmelin and Ostrea (Gryphaea) angulate of Lamarck*, in *The Conchological Miscellany*, 1854–1858.

Page 56. P. Gellé, *Café de Turin*, Nice, France, postcard, 1908.

Page 62. Auguste Feyen-Perrin, *Return of the Oyster Fishers*, 1908.

Page 64. Schell and Hogan, *The Oyster War in Chesapeake Bay* (detail), *Pirates Dredging at Night*, in *Harper's Weekly*, March 1, 1884.

Page 68. Carlo Ponti, *Oyster Vendor in Venice*, circa 1850.

Page 70. Eugène Boudin, *Le marché aux poissons*, 1875.

Page 73. Mason Jackson, *The First Day of Oysters*, in *The Illustrated London News*, vol. 39, 1861.

Page 76. John Fairburn, *Taking-in the Natives*, 1823.

Page 82. Berenice Abbott, *Oyster Houses, South St. & Pike Slip, Brooklyn*, April 1, 1937. Brooklyn Museum Collection, New York.

Page 83. Receipt for Oysters a la Rockefeller, Antoine's, New Orleans.

Page 84. © Tomales Bay Oyster Company. Used with permission.

Page 86. Anonymous, *The Lausten Brothers at Swan Oyster Depot*, San Francisco, 1934.

Page 88. J. Smith, *Oysters Fresh Every Day*, circa 1820.

Page 91. Primeval oyster fossil, *Umbrostrea emamii*, photograph by Manuela Schellenberger, © Bavarian State Collection of Paleontology and Geology, 2001.

Page 92. George Frederick Watts, *Tasting the First Oyster*, circa 1883.

Page 95. Stefan Claesson, *Shell Midden on Elizabeth Island*, 1888.

Page 97. Robert Bernard, *Huître. Ostrea*, in *Histoire naturelle, vers testacés à coquille bivalve irrégulière*, plate 180, 1772.

Page 100. Jean-François de Troy, *The Oyster Dinner*, 1735.

Page 102. *The Birth of Venus*, fresco in Casa della Venere in Conchiglia, Pompeii, circa first century C.E.

Page 106. Frans Floris, *The Feast of the Gods*, circa 1550.

Page 107. Dirck Hals, *Merry Company at Table*, 1627–1629.

Page 139. O.T. Olsen, *O. edulis*, in *The Piscatorial Atlas of the North Sea*, 1883.

Page 142. John Van Voorst, *Ostrea edulis* (detail), 1850.

Pages 148–61. Maps and illustrations © by Falk Nordmann, 2022.

INDEX

Note: Page numbers in italics refer to illustrations

ANDREAS AMMER, born in 1960 in Munich, works in television, teaches at university, and directs operas, but primarily he collaborates with musicians, and he has written many prize-winning radio plays.